HYDROINFORMATICS

Hydroinformatics

Information Technology and the Aquatic Environment

M.B. Abbott

Professor, International Institute for Hydraulic and Environmental Engineering, Delft, The Netherlands

Avebury Technical

Aldershot · Brookfield USA · Hong Kong · Singapore · Sydney

Published by
Avebury Technical
The Academic Publishing Group
Gower House
Croft Road
Aldershot
Hants GU11 3HR
England

Gower Publishing Company
Old Post Road
Brookfield
Vermont 05036
USA

A CIP catalogue record for this book
is available from the British Library.

ISBN 1 85628 832 3

Printed in Great Britain by
Billing & Sons Ltd, Worcester

Contents

Preface

It is in the very nature of that thinking activity that nowadays passes by the name of *research* that it can sometimes lead one into a very strange and unfamiliar place. As one proceeds to explore such a place, however, one meets others who have thought their way into these same new surroundings. With great difficulty — for everything is difficult here — one gets into conversation with these fellows in disorientation, and, slowly — for everything also proceeds slowly here, one compares notes about the paths followed to get to this place, and the different impressions that the place makes; and then, as is inevitable in human discourse, one gives the place a name. The name that has been given to one such place, where a considerable band of individuals has now assembled, is *hydroinformatics*.

However, it is just as much in the nature of *researchers*, that no sooner do they have a name and a vague idea than they presume to teach others about their new discovery. As one of the researchers-cum-teachers who have discovered hydroinformatics, and indeed as the one responsible for inventing this very name, it has fallen to my lot to describe what I have made out of this place called hydroinformatics and to indicate what we might reasonably expect from it in the future.

Just as inevitably in this respect, however, each of us who has arrived in this locality has proceeded along an own path, sustained by an own baggage. Some of us have arrived by one or another way of simulation modelling, and so have brought with us a baggage filled with mathematical models; others have travelled along the paths of data modelling, and so arrive loaded down with a variety of data bases; while others again have followed the routes of artificial intelligence, and most of these bring with them a baggage of knowledge-based-system technologies. Accordingly, depending upon the way that they have travelled and upon the baggage that has sustained them on their way, the new arrivals tend to see this place called hydroinformatics from different perspectives, and so with different emphases. For myself, the path has been that of simulation modelling and the baggage that of numerical modelling, and even now, despite some considerable effort, I have only an incomplete and mostly second-hand impression of the other paths that lead to hydroinformatics, and the quite other baggage that has sustained these, my colleagues, as they have travelled along the other paths. However, since all of these paths have their own relevance to hydroinformatics, even as they converge from such widely-differing points of the compass, I have no choice but to try to describe these ways, and to take account of the baggage that has been brought to hydroinformatics along them. This is a great presumption on my part, of course, but one that appears to be unavoidable.

This work is written both for those who have already arrived in the domain of hydroinformatics and for those who have not. For these last, a few words of explanation must be given at once. It must first be explained that *hydroinformatics has to do primarily with the social dimensions of hydraulics and environmental engineering*, and that it proceeds, in particular, within our currently-predominant, social-economic environment of information technology. It is for the most part concerned with *how* science is applied, rather than with science itself.

This work is, accordingly, not primarily a scientific work. Indeed, the very fact that it proceeds in social dimensions means that it has to take account of *non-scientific* aspects. These arise already when we consider how information technology is influencing the way that we have come nowadays to see and to think about the aquatic, including marine, environment. These non-scientific aspects are further accentuated as we introduce common-sense truths, as facts and rules, into our systems, whereby what is 'common-sense' to one social group may be something altogether different to another social group. The social arrangements which are introduced in order to accommodate these 'truths', and which become institutionalised through legislation, regulation, contract and usage, are so important in environmental studies that hydroinformatics must take proper account of them.

The non-scientific aspects return in another form again when we extend our work into ecology. So far, most ecological studies have been directed towards numerical simulations of simple extensive variables, such as species biomass and energy exchanges, as expressed mathematically in systems of simultaneous ordinary differential equations. We shall call this approach the 'Lotka-Volterra paradigm', after the names of its initiators. However, it is inevitable that methods of artificial intelligence will come to be applied increasingly here, as explained in the body of the work. However, the 'intelligent agents' that we must introduce in order to represent intelligent creatures can only be nominally scientific, in that all we can do in this way is to infer averaged phenotypal behaviour on the basis of our own limited and conditional beliefs about the psychic states and consequent behaviours of the corresponding genotypes. The aspect of uncertainty in our own beliefs becomes more pronounced again in further human-social, or sociological studies, which increasingly commonly accompany investigations into water-resources projects. As we proceed further to introduce the effects of uncertain inference into our descriptions, we necessarily enter into ways to limit the influences of our own preconceptions. As is explained in the work itself, this ultimately necessitates the use of subsymbolic paradigms such as are currently being realised through the use of neural networks and their emulators.

From these considerations alone — and more are introduced in the body of the work itself — it follows that a book about hydroinformatics must span a space in thought between various branches of natural science and very much that is not comprehended by natural science. Hydroinformatics is thus, in a very definite sense that will be explicated in this book, a *technology*. It accordingly opens up in a particularly acute form the whole question concerning the relation between natural science and technology, which is to say that it poses, to use the title of Martin Heidegger's seminal essay on this subject, *The Question Concerning Technology*.

This most basic and fundamental question is pursued within the present context, of applying information technology to the problems of the aquatic environment, at two levels. The first of these is the level that we can best describe as 'the philosophical'. The need to accommodate the non-scientific aspects here leads us to place the usual 'philosophers of science', such as Frege, Mach and Carnap, largely to one side. We find that we have to step back, at the very least to the phenomenology of Edmund Husserl, but we then find ourselves propelled, apparently by sheer necessity, further, to Heidegger. As we shall see, however, it is only by taking such an approach that we can provide ourselves with even a minimal conceptual foundation upon which to pursue a serious discussion of this subject. Although many readers may not be familiar with the conceptual apparatus used here, its introduction within the context of practical examples should make it quite accessible. For those with a background in philosophy, on the other hand, this aspect of the present work, although quite conventional, may serve to illustrate the extraordinary power of the Heideggerian conceptual apparatus.

The progress of the work itself shows, however, that even this philosophical level will not suffice to bring us to a depth of understanding that is truly satisfactory. Accordingly, as the book progresses, it moves by degrees to another level, which can then best be described as 'the theological'. This is a level that is rarely if ever entered in technological studies, but again we are here led naturally and inevitably into it. In this case also the approach followed is fairly conventional by present-day standards, roughly following a trajectory that passes through Søren Kierkegaard and Karl Barth. It becomes clear as we progress that it is only at this level that we can establish a proper conceptual foundation for hydroinformatics, as indeed for technology generally. Once again, the way of thinking may be unfamiliar to many readers, but the context should suffice to make its application and relevance clear enough. On the other hand, for those who are familiar with present-day, 'post-modernist' theological thinking, this work may serve as a particularly transparent example of the application of this kind of thinking.

In this work, as in technology generally and hydroinformatics in particular, the scientific and non-scientific aspects are closely intertwined. However, by proceeding to these 'philosophical' and 'theological' levels, it becomes possible to elucidate the subject in its full depth, passing from the established natural-scientific layers, through the pre-scientific layers, as uncovered in particular by Husserl, thence through the level of the essential Being of the objects displayed, as illuminated so extensively by Heidegger, and finally arriving at the conceptual bedrock of the subject. As we arrive at this basal stratum, we are of course necessarily far removed from the usual thought world of natural science, with its conventional *logos*. We are here in the realm where discourse can only proceed at all in allegorical, including mythological, terms, so that this is necessarily the realm of *mythos*. Indeed, the main thread that is drawn through this book, and which is progressively tightened as the work progresses, is that of 'the number myth', as this presents itself in all manner of different forms, or metaphors. This is presented as the basic mythological structure which underlies hydroinformatics and which hydroinformatics itself connects to the task of caring for the waters that feed and cleanse our planet, which are, to use Leonardo da Vinci's simile, the 'arteries and veins' of the biosphere. At this level, the work rather inevitably draws heavily on the psychology of Carl Jung and various authors in the field of anthropology, and in particular Claude Lévi-Strauss; but it has also been strongly influenced by the philosophical views of Karl Jaspers and the theological considerations of Rudolf Bultmann, as recorded in their celebrated dispute over the role of mythology in present-day society. The reader coming from natural science may have met some of these notions elsewhere, and notably in quantum physics, biology and astronomy, but for most readers they will probably appear rather novel. In order to provide some support along this path, a number of footnotes have been inserted in which the most common of the notions are introduced. This aspect may also be of interest to anthropologists, philologists, various sociologists and the many others who are concerned with the conservation, reification and propagation of myths and mythological language.

As concerns the division of the presentation itself, this is divided into three chapters, supplemented, in imitation of Kierkegaard, by a 'concluding unscientific postcript'. The first chapter is largely scientific and quite pragmatic, scarcely proceeding at all beyond the 'philosophical' level. The second chapter expands the subject out into its social dimension, which necessarily takes the discussion from the 'philosophical' level to the 'theological' one. In the third chapter, the discussion alternates between these two levels, while connecting them to the most obviously relevant scientific developments. The postscript proceeds entirely at the 'theological' level.

As the work is intended for a much wider audience than engineers and applied natural scientists, mathematical formulations have been relegated to one section, which concludes the third chapter.

The question that must arise even after reading only this preface is: what is this book *really* about?; and to this I can only answer in the most general terms, that it is about that place in thought where the waters of our world and our current informational revolution come together. Correspondingly, it is my hope that it will be of interest and value to everyone who genuinely cares about this place in thought ... where the waters of the world and our current informational revolution come together.

THE HAGUE, March 1991.

Acknowledgement: I thank the many colleagues who read the original draft and the various stages in its elaboration. I also thank the Danish Hydraulic Institute (DHI) for providing all the illustrations and much descriptive material besides. My appreciation is extended to the editors for their exceptionally thorough work.

Thus the coming-to-presence of technology harbours in itself what we least suspect, the possible arising of the saving power.

... The closer we come to the danger, the more brightly do the ways into the saving power begin to shine and the more questioning we become. For questioning is the piety of thought.

Martin Heidegger, *The Question Concerning Technology*

1 The nature and origins of hydroinformatics

Because the essence of modern technology lies in Enframing [*Gestell*: the assembling and ordering of our physical world within a rational framework], modern technology must employ exact physical science. Through so doing, the deceptive illusion arises that modern technology is applied physical science. This illusion can maintain itself only so long as neither the essential origin of modern science nor indeed the essence of modern technology is adequately discovered through questioning.

Martin Heidegger, *The Question Concerning Technology*

1.1   Towards a characterisation of hydroinformatics: the first cycle

Hydroinformatics is a new name, being the name of a new possibility. We shall for the most part consider what we are to understand by this new name, and only later will we return to the new possibility which hydroinformatics provides.

Hydroinformatics evidently has to do with water and information. It has, however, to do with water and information as these are associated in one particular kind of relation. We can characterise this kind of relation as one that is *scientific in the modern sense*. This is to say, as a first approximation, that it is information that is orderable, countable and subject to computation (Heidegger, 1977, pp. 17-23). Thus the kind of information that enters into hydroinformatics can be given a definite representation in the 'information theory' of modern science, as that which resolves a state of uncertainty, and indeed one bit of information is there conventionally defined as that which resolves a yes or no, or an on or off or a 0 or 1 or whatever other binary state of uncertainty. We then conceive of a situation in which a given water body is described by a certain body of such 'standard-scientific' information. Indeed, in hydroinformatics, the water body, originating as an object of our sense-perceptions, is 'replaced' by a body of information which in its most basic representation consists only of strings of binary signs. Such a view of the waters of the world must at first appear as a very *idealised* one, and, in turn, a very limited, circumscribed and deprived one. In the words of Heidegger (1927/1962): 'This manner of knowing [nature] has the character of depriving the world of its worldhood in a definite way.' What we experience in this impression is that in hydroinformatics we appear to lose completely the *immediate visual impact* of the painter — the seaborne mercantile optimism of a Van der Velde, the cold watery death of a Géricault, the shimmering haze of a Turner, the sparkling freshness of a Seurat, or, for that matter, the dream-like sea of a Magritte. Here we appear to have none of the *immediate aural-visual impact* of the poet — of a Hölderlin or a Tennyson, or of the composer — a Mendelssohn, a Wagner or a Debussy. On the other hand, we know that any painting can be scanned such that its pigments and their distributions over the canvas can be represented in a digitised form. Similarly the words of the poet can be digitised simply by keying them into any word-processing system, while similar, albeit more specialised systems are available for musical notations as well. Clearly, then, the special features of hydroinformatics — which features we experience as idealised, limited, circumscribed and deprived in the standard sense of modern science and which were identified as such already by Heidegger (and, before him, by Husserl) — cannot reside simply in the possibility of reducing them to a minimal, digitised form.

A moment's reflection shows that it is not the *information itself* that strikes us as idealised and deprived, but it is the *knowledge that this information conveys to us*. Hydroinformatics imparts one kind of knowledge; the examples from the arts provide another kind. What, then, is the difference between these two kinds of knowledge? Let us first follow the later Heidegger, the Heidegger of *The Question Concerning Technology* (pp. 16, 17):

> The hydroelectric plant is set into the current of the Rhine. It sets the Rhine to supplying its hydraulic pressure, which then sets the turbines turning. This turning sets those machines in motion whose thrust sets going the electric current for which the long-distance power station and its network of cables are set up to dispatch electricity. In the context of the interlocking processes pertaining to the orderly disposition of electrical energy, even the Rhine itself appears as something at our command. The hydroelectric plant is not built into the Rhine River as was the old wooden bridge that joined bank with bank for hundreds of years. Rather the river is dammed up into the power plant. What the river is now, namely, a water power

supplier, derives from out of the essence of the power station. In order that we may even remotely consider the monstrousness that reigns here, let us ponder for a moment the contrast that speaks out of the titles, 'The Rhine' as dammed up into the power works, and 'The Rhine' as uttered out of the art work, in Hölderlin's hymn by that name.

... Everywhere everything is ordered to stand by, to be immediately at hand, indeed to stand there just so that it may be on call for a further ordering. Whatever is ordered about in this way has its own standing. We call it the standing-reserve [*Bestand*]. The word expresses here something more, and something more essential, than mere stock. The name standing-reserve assumes the rank of an inclusive rubric.

... The fact that now, wherever we try to point to modern technology ... the words setting-upon, ordering, and standing-reserve, obtrude and accumulate in a dry, monotonous, and therefore oppressive way, has its basis in what is now coming to utterance.

What Heidegger explained more generally in his epochal essay was that modern technology, through its alliance with modern science, reduces the whole world of nature to a standing reserve, by regulating and securing, and that this again is achieved through ordering, counting and calculating. Hydroinformatics has to do with knowledge of the waters as standing reserves.

We are here rather closer to the essence of what it is that separates off and demarcates hydroinformatics, and what it is that impresses upon us its scientific characteristics of idealisation and deprivation.

In order to proceed further in this first characterisation of hydroinformatics, let us revert to the earlier Heidegger, of *On the Essence of Truth*, of 1930 (see Biemel, 1973/1977). We are then led to contrast the concept of a *truth of modern science*, as represented here by hydroinformatics, with the *truth of a work of art*. The first kind of concept is usually expressed in terms of a correspondence between two 'statements', which we can now already represent as a correspondence between two strings of binary digits. When seen within this frame of reference, a certain correspondence between a segment on one string and a segment on another can be said to establish a concordance, which we interpret as truth.

However, this view of truth as correspondence itself dates from the scholastics; it is itself derived in a direct line from Christian philosophy:

> *Veritas* always means in its essence: *convenientia*, the accord of 'what-is' itself, as created, with the Creator, in accordance with the destiny of the creative order.

Now, however:

> The creative order as conceived by theology is supplanted by the possibility of planning everything with the aid of worldly reason, which is a law unto itself and can claim that its workings ... are immediately intelligible.

Then, as Biemel comments on this theological origin of scientific truth, (1973/1977, p. 79):

> Even where this position is abandoned, the interpretation of truth as the correctness of correspondence survives, acquiring a quasi-absolute validity, and it is forgotten how this interpretation was originally justified.

It was against this 'standard-scientific' view of truth that Heidegger advanced another view, more suited to the comprehension of artistic truth. This is the truth that is experienced as *revelation*, and so as a revealing: an unconcealment which is simultaneously an attunement. This kind of truth is received only in a condition of openness to that which is revealed, or which becomes unconcealed in order that the attunement may occur. This kind of truth is experience, so to say, as a 'beauty in the eyes of the beholder'. We shall return later to the question of the relation of scientific truth to this kind of 'aesthetic' truth within the ambit of technology.

There is, however, a further concept of truth which is considered by Heidegger, which is that of *common sense*. This is 'the actual truth that can give us a standard or a yardstick'. Common sense associates truth with efficacity and utility; its experience is always touched with an element of intentionality. Thus, for example, the truth about the river may well be one common-sense truth for the fisherman, corresponding to his fishing interests, and a quite other common-sense truth for the farmer, corresponding to his farming interests. Here we have to do with a truth that is not a scientific truth, even as it is also not an artistic (or aesthetic) truth. We shall subsequently argue that, when viewed as a social benefit, hydroinformatics has to take account of common-sense truths. Correspondingly, to the extent that such truths are formalised and systematised in laws, contracts and other social arrangements, hydroinformatics has to comprehend these forms and systems. Very often, common-sense truths come to appear in hydroinformatics in the form of *heuristics*: common-sense knowledge then commonly takes the form of the validity of certain heuristic rules. (See Barr and Feigenbaum, 1981, I, pp. 28-30; also Nilsson, 1980, p. 72). Common-sense knowledge is, of course, the very stuff of everyday activity: it determines how things are designed, how they are constructed, and how they are managed.

In fact however, and quite generally, the function of knowledge is to *discriminate* between truth and untruth. In this sense, however, knowledge is an attribute that is unique to the biological world. Thus, although we can *describe* elements of knowledge in the form of information about rules, so that this in its turn acts upon information about facts so as to 'descry truth', knowledge *per se* cannot be *replaced* by information: it has no informational form, in the standard-scientific, information-theoretic sense.

(Of course, above knowledge we have wisdom again, but that is way beyond our present purview! Indeed: above information is knowledge and above knowledge is wisdom; the scientific content decreases rapidly as one proceeds upwards).

Thus, to conclude this first cycle along the way to a characterisation of hydroinformatics, we may see this subject as one that treats all scientifically-defined information and information flows associated with a water body in such a way that this body is regarded as a standing reserve for social use, so that it is subject to the interested actions of common-sense heuristic rules and facts and their associated reasoning, as well as to all the required actions of ordering, counting and computing.

If this was all that had to be said of hydroinformatics then we could just as well consign it to the place of the other 'dismal sciences' in which 'the words setting-upon, ordering, and standing-reserve, obtrude and accumulate in a dry, monotonous, and therefore oppressive way'. There is, however, much more to it than this.

1.2   Further towards a characterisation of hydroinformatics: the second cycle

Hydroinformatics is concerned with the information flows that accompany and govern the flows of fluids and of all that these fluids transport. However, even if only because common-sense truths have also to be accommodated within hydroinformatics, this subject cannot itself be a science, even though it draws upon many sciences: *it is, in a very essential sense, a technology*. Let us first be very clear about what we mean by this, reverting to the Heidegger of *The Question* (1977, pp. 12,13):

Technology is therefore no mere means. Technology is a way
of revealing. If we give head to this, then another whole realm
of the essence of technology will open itself up to us. It is the
realm of revealing, i.e. of truth.

This prospect strikes us as strange. Indeed, it should do so,
should do so as persistently as possible and with so much
urgency that we will finally take seriously the simple question
of what the name 'technology' means. The word stems from the
Greek. *Technikon* means that which belongs to *technē*. We must
observe two things with respect to the meaning of this word.
One is that *technē* is the name not only for the activities and
skills of the craftsman, but also for the arts of the mind and the
fine arts. *Technē* belongs to bringing-forth, to *poiēsis*; it is
something poietic [creative, formative, productive, active].

The other point that we should observe with regard to *technē* is
even more important. From earliest times until Plato the word
technē is linked with the word *epistēmē*. Both words are names
for knowing in the widest sense. They mean to be entirely at
home in something, to understand and be expert in it. Such
knowing provides an opening up ...

Technology is a mode of revealing. Technology comes to
presence in the realm where revealing and unconcealment take
place, where *alētheia,* truth, happens.

To draw upon other expressions of Heidegger, we might speak of a
'coming-to-presence', or of a 'surfacing', of certain of our innermost but
collective thoughts and desires, as material objects in our outer world of name
and form. If we lean over backwards to provide a popular formulation, we
might describe the essence of technology as that of *making dreams come true.*[1]

---

[1]There still remains a remarkable amount of confusion about this, and especially about the
relation between science and technology. Technology has to do with *creation in our material
world*: of transforming both our perceived and our still-unperceived aspirations and desires into
physical realities. Technology is mankind's extension of natural creation. The relation between
these kinds of creation is, at its best, expressed in the celebrated words of Goethe (see also
Dilthey, 1976, p. 60): 'Just as in the sphere of morality we are meant to approach the highest
good so, in the intellectual sphere, contemplation of an ever-creative nature is intended to make
us worthy of participating in its production.' Science, on the other hand, has to do with
understanding our world, whether this be our outer world of name and form (as in natural
science) or our inner world of soul and spirit (as in dogmatic science). Science frequently
contributes to technology, but then so do aesthetics (as in 'industrial design'), business, and many

Hydroinformatics also conforms to this view of technology when viewed in relation to the flow of fluids, and in the first instance to the flow of water. We hardly need reminding here that water is the first requirement of agriculture; that it is this that transports sediments, which include those nutrients which promote the productivity of agriculture; and it is primarily water that removes accumulated wastes, which otherwise would cause illness and disease. It is the flow of the waters that underlies — feeds and drains — the entire trophic engine of the biosphere. To use a simile beloved of Leonardo da Vinci, we can describe the courses of the waters as the very arteries and veins of the biosphere. Hydroinformatics, then, also corresponds to this Heideggerian concept of technology applied to these arteries and veins of the biosphere.

Now, however, in and through the whole development towards hydroinformatics, something very special is happening to the way that technology operates in relation to the waters, to the way that revealing and unconcealment takes place in the case of the waters, to the 'where' of 'where *alêtheia*, truth, happens'. This something special in the way that technology comes to operate, its 'where-on-the way', has itself to be revealed, unconcealed, or brought to presence.

As will be set out in some detail in the next section, but as will already be clear from everyday experience, this 'something special' has to do with the ubiquitous application of *numerical modelling*, whereby it might be said that 'truth happens almost always nowadays in numerical models'. This 'where' is, superficially and initially at least, 'within the numerical model'. But what is it that is really happening already in our current, almost-obsessive application of numerical modelling to everything that flows? What is it, *essentially*, that heralds the approach of what has already been introduced here as hydro-informatics, in its development through numerical modelling? We can perhaps best answer these questions through some illustrative examples.

The first of these examples is that of applications of simulation modelling to the design and operation of storm-sewer systems (SSS's), so of applications within an industry sector which invests several tens of billions of dollars a year world-wide. A typical layout of an SSS for part of a small town is shown in Figure 1.

---

other activities besides, as will shortly be explained in relation to hydroinformatics. Similarly, technology often contributes to science. But to ask of technology that it should base itself entirely upon science is every bit as misguided as to ask of science that it should devote itself exclusively to the interests of technology. Apart from anything else, who could ever predict the futures of such bases and interests correctly? An astrologer perhaps? In point of fact, in the world as it actually functions, technology and science proceed along paths that are, so to say, orthogonal, the one to the other. Correspondingly, each has its own rights and its own responsibilities.

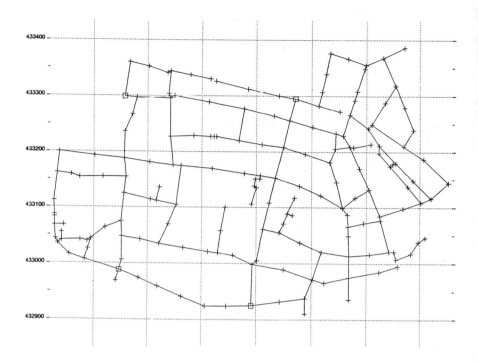

Note: The network for part of a small town is shown. The information contained here is greatly augmented by details of the shapes and sizes of pipes, culverts and manholes, invert levels, positions and characteristics of pumps, weirs and other control elements, the control system, descriptions of operating rules and practices, hydrologic parameters over the areas supplying water, meteorological data and, increasingly, water-quality data. Almost all of this is in number form.

Figure 1   A typical storm-sewer network

This illustrates only the lines of pipes and culverts and the position of manholes in the system, but to this information must be conjoined the shapes, dimensions and levels of the pipes and culverts, the dimensions of the manholes and the invert levels of the pipes and culverts that intersect there, the areas and hydrologic characteristics of the areas collecting storm water and delivering it to the system, and many other data besides, almost all of them expressed in the form of numbers. Prior to the use of large-scale hydraulic simulation, this information was available on drawings and in tables, filed away in local archives and only tracked down and taken up for use as and when some further employment could be found for it. This information was thus *available but not readily accessible.* However, with the development of hydraulic simulation modelling, and so through the increasing intercession of hydraulics knowledge as will be traced in the next chapter, all of this information becomes *immediately* accessible: it is available to guide diagnoses of causes of surface flooding and other calamities, for the design of ameliorating measures, for on-line control, for reconfiguring and redesigning the system, for setting new operating rules, for introducing new bye-laws, for settling insurance and legal disputes and for many other purposes besides. This is to say that information that had earlier ceased to act upon our present world now enters this world as an *agent of causality*. We might then try to express the essence of this new situation by saying that information that was, so to say, *asleep*, now becomes *awake*. In effect, this information, which is composed almost completely of numbers, *wakes up into our world* to become an agent of causality in this world, and it does this for the first time in modern history. We can place this within a current metaphysical framework by introducing the French verb *relever*, literally 'to lift up', as popularised in our own times by Derrida. (See 1972/1982, especially p. 20; Derrida introduced *relever*, as a translation of Hegel's *Aufheben*, to indicate an act of lifting up in which is inscribed an effect of substitution and difference, in effect of *supersession*). The numbers concerned are 'awakened' and so 'return to life' through the agencies of our hydraulics knowledge and our digital machines, so that we can say of them that *ils se relèvent, ils se réveillent à la vie.*[2]

In the language of Heidegger (1927/1962) this is however to say something more again: it is to say that these numbers have not only acquired another presence-at-hand (*Vorhandenheit*), but they have attained to another kind of Being-in-the-world, another kind of *In-der-Welt-sein*; indeed, as we shall subsequently see, this is a most *basic kind* of Being-in-the-world, *eine Grundart des In-der-Welt-seins*. In order to make this point most emphatically, we can place it within its immediate social context and say, provocatively but with serious intent, that:

The numbers have begun to function in another way.

Our next example is taken from hydrologic modelling, where simulation systems like the European Hydrologic System — *Système Hydrologique Européen* (SHE: see Abbott *et al*, 1985) — permit the deterministic modelling of the land phase of the hydrologic cycle for complete catchments. The system is schematised in Figure 2a. Some typical sets of descriptive distributions that are used to set up a particular model are shown in Figures 2b and 2c, while one

---

[2]The second expression was suggested by Professor J.A. Cunge as a natural complement of the first.

output distribution is shown in Figure 2d. What we observe here is how such a model *brings together* sets of numbers that pertain to aspects of the material world that are normally considered to be quite widely separated, and indeed such that one set may be associated with one science (*e.g.* geology) while another may be associated with a quite other science (*e.g.* plant physiology). Quite symmetrically, the possibility of associating different kinds of scientific knowledge through such models provides a new incentive to bring certain branches of science further into the domain of the ordered, the numbered and the calculated. In this way, almost everything that we know about such a physical region comes to be expressed in terms of numbers and to be integrated through and by numbers and operations on numbers. Thus almost all our knowledge is *brought together* in our present world, as an instrument of causality in this world, through the action of the numbers, and this bringing together has again no modern-historical precedent. Once again the numbers attain to another Being-in-the-world, another *In-der-Welt-sein*. We thus see from another point of view how 'the numbers have begun to function in another way'.

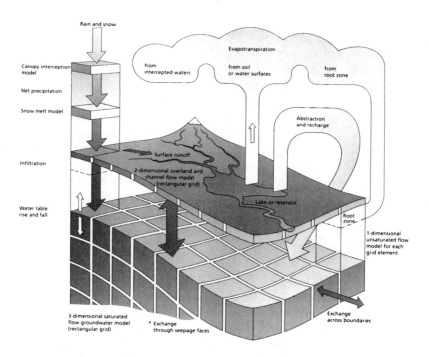

Note: Schematisation of the European Hydrologic System — Système Hydrologique Européen (SHE) — as used for distributed, deterministic hydrologic modelling.

Figure 2a   Overview of the SHE

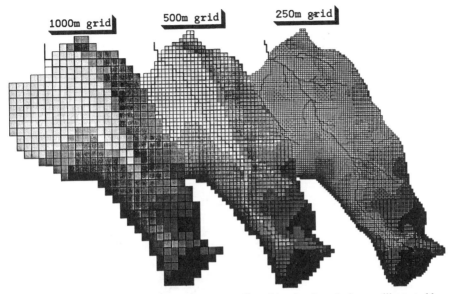

Note: Such a system allows for the use of any number of levels of resolution, as illustrated here for the description of the surface levels, or topography of a particular basin.

Figure 2b   Typical input distributions used in the SHE, 1

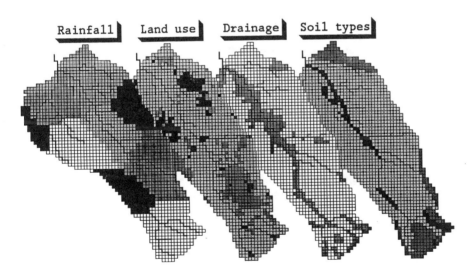

Note: Typical distributions of parameters used in setting up the model.

Figure 2c   Typical input distributions used in the SHE, 2

Water-levels in river and on the soil surface

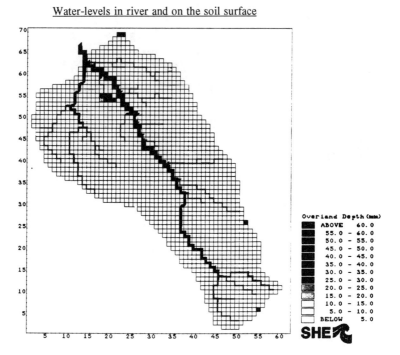

Note: A typical instantaneous, space-distributed output of the system. Outputs may also of course be taken as time variations at particular points and may describe water-quality indicators as well as flow quantities. Illustrations of this kind are invariably given in colour, greatly enhancing their legibility, but this facility was unfortunately not available for this publication.

Figure 2d   Typical output distributions used in the SHE

The third example hinges upon model results of short-period wave action on an approach channel and lagoon entrance, as illustrated in Figure 3.

The point to be brought out here is that, through this computerisation of hydraulics knowledge, complex phenomena that were never before associated with numbers in this way now come to be described by sets of numbers. Moreover, these sets of numbers are themselves very extensive: the computation of one particular wave field for a single lagoon-entrance, as shown in Figure 3, provides about a billion seven-decimal-digit numbers as output alone. Since we can print about 25,000 such numbers on a square metre of paper, the output of this particular simulation alone would fill some 40,000 square metres — enough to embalm, Christo style, l'Etoile or Trafalgar Square or almost any other central city concourse. Here it is the *complete phenomenon* which 'awakens' into our immediate world for the first time through the agency of the numbers, while these numbers have here as well another kind of Being-in-the-world, another *In-der-Welt-sein*. We see from yet another point of view how 'the numbers have begun to function in another way'.

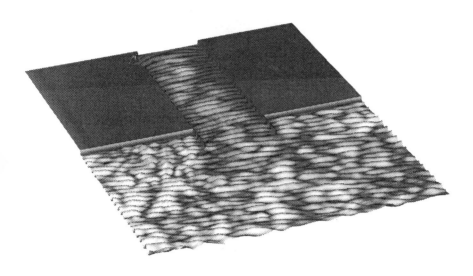

Note: An example of a short-period-wave field produced by waves with both frequency and directional spreading at one of the entrances of the Venice lagoon.

Figure 3   A short-period-wave field

Just one more example! Figure 4 shows a ship in a yawing motion at a single-point mooring, in which the ship's motion and its wave field are coupled together in one model. Through the use of more modern computer-graphic facilities, we can have here (when colour is available) a quite highly realistic visual impression of a ship and its surrounding waves, whereas 'in reality' there *is* only a very extended set of numbers. We see how in this case the numbers are 'brought to life' with a new immediacy in our world, and again as a means of influencing our immediate actions in the world, and this capability is clearly also without modern-historical precedent. Through these means the numbers attain to yet another Being-in-the-world, to yet another *In-der-Welt-sein*. In this way also, 'the numbers have begun to function in another way'.

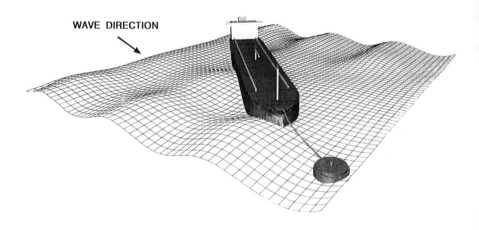

Note: A ship moored at a single point is subjected to waves from the beam in order to provide the wave fields around the ship (and the ship's motion if necessary). It is usual in illustrations of this kind to use dense natural colours so as to provide an almost photographic effect. Once again, and regretfully, these facilities were not available to this publication.

Figure 4   A ship moored in a beam sea

Thus, the first, rather everyday observation, that truth mostly happens nowadays in numerical models, leads already from several points of view to the distinctly controversial proposition that, to repeat ourselves once yet again: the numbers have begun to function in another way.

It will moreover now be clear that by this it is in no way meant that the numbers have begun to function in another way in any scientific sense — whether in the standard-mathematical sense of the ancient Greeks, of Weierstrasse and Hilbert, or in the otherwise-scientific, non-standard-mathematical senses of Leibnitz and Robinson — but in a *technological sense that is not scientific*. It is then seen that this other way in which the numbers are beginning to function has little or nothing to do with science-as-such at all, but it clearly has quite a lot to do with how science is applied in our outer world of name and form. Correspondingly, research in this 'other way of functioning of the numbers', which is of the essence of research in hydroinformatics, *is not scientific research*, even though it can be construed as research upon how science is applied, or how it participates in the 'bringing forth' that is technology. In this way, as 'research into revealing', hydroinformatics research belongs very essentially to research in technology.

This research is, then, concerned with the 'where' and the 'how' of the coming to presence and the bringing forth of this other way of functioning of the numbers in our outer world. It has to do, at least initially, with how works are currently designed and how they might better be designed in the future, with how these works are currently realised and how they might better be realised in the future, with how complete systems are currently managed and how they might better be managed in the future, and so on. Although it must always start from where those that it serves, its clients, are presently situated, it must also always seek to lead these that it serves to another, higher place. As we shall see, one of the places where this process begins is in simulation using numerical models, even as it already proceeds much further, into areas that have apparently very little to do with simulation using numerical models.

We may already here make a more general, even though still tenuous, link between this other and in some ways deeper view of the development of hydroinformatics, and the view introduced in the first cycle, by observing that, through the practice of numerical modelling using digital computers, the numbers are already being used for all sorts of *purposes* that were scarcely even envisaged only a few decades ago. As a result, these new ways of using numbers, in numerical models, start to have their own impacts on common-sense truths, as these operate at the design, construction and management stages of hydraulics works, and they thereby have an impact upon our whole view of the intention, purpose, efficacity, utility and other social aims of these works. Then, as a first, rough, superficial and simplistic way of seeing the new way of functioning of the numbers, we can point to the fact that the numbers are already having an entirely new impact on human interventions in the world of nature. For us, as hydraulicians, this is a new impact upon the very arteries and veins of the biosphere.

Thus, from the point of view of this second cycle, hydroinformatics is the coming to presence in water technology of a new way of functioning of the numbers; it is a new way of creativity, of formation, of production, of activity in relation to the waters, realised through the intervention of the numbers; it is a new place where truth happens, between the waters and the numbers.

1.3   From computational hydraulics to hydroinformatics

1.3.1   The origins of computational hydraulics

It is now twenty years, almost to this day of writing, since the name of *Computational Hydraulics* was introduced by the author to describe a hydraulics that was formulated in such a way as to suit it specifically to the ways of working of digital computers (*e.g.* Abbott, 1979). The notion of 'the computational sciences', generally, was based upon the observation that science as a whole, prior to our current 'age of computers', had been formulated in such a way as to suit it to the most efficient ways of working of our own mental capabilities. The constraints thereby placed upon the development of science were scarcely seen as such by the practitioners of this science, for, after all, there appeared at that time to be no other way of doing things. Of course, a Leibnitz, a Babbage or an L.J. Richardson would occasionally remind science of what it might accomplish if it embarked upon its own 'industrial revolution' and acquired tools, like other trades. Such views were, however, regarded as, at best, aberrant, and at worst downright subversive to the true purpose of science, which was seen as that of giving free play to the powers of the unaided intellect.

D.R. Hartree, one of the great pioneers of numerical methods (*e.g.* Hartree, 1952), once observed that these methods were regarded with disdain, bordering on contempt, by most of the scientists of his time, who saw in the writing of beautiful analytic functions and in the elucidation of the relations between these a 'superior calling' — and certainly one endowed with a much higher prestige in academic circles. At the time that Computational Hydraulics was proposed as a viable field of study, the 'higher reaches' of hydraulics were those that merged into a fluid mechanics that corresponded entirely to the analytical-mathematical paradigm. Even in hydraulics, the most prestigious achievement in academic circles at that time was to publish one's work in the *Journal of Fluid Mechanics*, which was in those days given over entirely to the analytical approach.

The early evolution of computational hydraulics took place against this background of a long-established mathematical-scientific tradition, with its own achievements and its correspondingly-established institutional arrangements. Computational hydraulics had to do with applications, but then it had at once to contend with the situation whereby the tradition had long ago laid down its own norms as to what constituted applications, to which it gave the generic name of 'applied mathematics'. Applied mathematics was the very stuff of this tradition. It was set out in very substantial texts, such as Lamb's *Hydrodynamics* (1931), and of course it filled out the scientific journals. In this situation, computational hydraulics, in its infancy, appeared distinctly poor, with few friends in the camp of the established mathematical sciences.

It had few friends, but it did have a few. In 1957, R.D. Richtmyer published his *Difference Methods for Initial Value Problems*, which summarised work that had been started at Los Alamos as a part of the first nuclear-weapons programme, and which had been continued at an academic level, in the first place at New York University (subsequently the Courant Institute). This work pointed to a mathematical foundation for computational fluid dynamics, and with this computational hydraulics, in the form of *functional analysis* (*e.g.* Riesz and Nagy, 1955; Kolmogorov and Fomin, 1957; Liusternik and Sobolev, 1961: see, more generally, Roman, 1975). Now it is of considerable interest, in retrospect, that, by the standards of the established tradition, this was a comparatively new branch of mathematics (in that some parts of it had only been elaborated in the twentieth century), so that little or nothing of it appeared in the standard texts on 'applied mathematics' (*e.g.* Milne Thomsen, 1955; or even Whitham, 1974). Several of those working in computational hydraulics, however, found in the formalism of functional analysis a useful way of formulating problems in ways that were more 'human-friendly', and thus connected to certain fragments from the traditional approach (such as differential equations), while still providing the means to proceed further, to translate these formulations into more 'machine-friendly' procedures. Over and above this, however, functional analysis provided a language and an overall conceptual structure for describing 'what the machine was doing' in ways that we mere mortals could understand. Thus it came about that practitioners in computational hydraulics learnt to think and to talk in terms of metric spaces and contraction mappings, leading on to normed linear spaces and their completeness conditions (to provide Banach spaces), and from this viewpoint they were able to use certain key notions of the theory of Hilbert spaces. From 1966 onwards, this approach provided the principal mathematical apparatus of computational hydraulics, and indeed for much of computational fluid dynamics also.

1.3.2   Computational hydraulics and the first three generations of modelling

Computational hydraulics was, however, in the first place a way of studying real-world problems. Its purpose was to support *numerical modelling*. Its utility proceeded from its ability to provide criteria for the success or otherwise of numerical models, whether predictively or diagnostically. Computational hydraulics developed as the *science* of numerical modelling of hydraulic phenomena. Like all other branches of modern science, it was essentially a mathematical science, but then one that used a mathematical apparatus which was already then rather far removed from that normally associated with applied mathematics. The development of computational hydraulics took place interactively with the development of modelling, with the interaction itself being strongly conditioned by the chosen mathematical formalism (Abbott, 1979).

It is common nowadays, when tracing the development of modelling, to distinguish between its different *generations*. The first of these is associated with the programming of computers to do much the same things as human beings do without computers: finding the roots of equations by standard, human-friendly methods, calculating values of functions, etc. The numerical side was only concerned at this stage with simplifying or speeding-up the established procedures, while in principle these procedure kept their human-friendly forms.

Some of the early users of these machines went beyond this stage, however, by observing that a procedure that was 'human-friendly' was usually very sub-optimal when compared with more specifically 'machine-friendly' methods. At the same time, the machine-friendly methods were usually quite distinctly human-unfriendly — which was where computational hydraulics came in, with its 'arbitrating' role. Some of the most machine-friendly methods were well adapted to the building of models that described variations of fluid properties in space and time: the so-called 'initial-value problems' of the traditional view. The most common of these methods, at least initially, were the so-called 'finite-difference' methods, in which a differential-equation description of the behaviour of a physical system could be approximated by a difference-equation description (Richtmyer, 1957).

Over the first decade of modelling, which extended roughly from 1960 to 1970, this approach was realised in the form of one-off, custom-built models, of which a particular masterpiece was the model of the Mekong Delta built in the mid-1960s. This stage in the development of modelling is usually referred to as the second generation.

At the institutional level, the second generation of modelling established the paradigm of the 'modelling group': a group or team of individuals, often of diverse backgrounds and talents, who together constructed models to special order. A considerable accumulation of understanding then occurred within this modelling group through the practice of its modelling work. However, this accumulation occurred for much the greater part within the minds and in the notes of the individual modellers: it did not get, as we say, 'fixed at the level of the code'. The institutional paradigm for second-generation modelling was established by the French consulting company SOGREAH from about 1956 onwards (Abbott and Cunge, 1981).

The clients for second-generation modelling services were themselves few in number, for such persons had themselves to have a considerable 'feeling' for numerical modelling. After all, these clients had themselves to be able to use the results obtained from a particular model to such a good effect as to cover the quite considerable costs of this custom-built model. If we should try to identify what *essentially* the modelling group and its clients already then had in common, we could perhaps do best to describe it as a shared *taste for* numerical modelling. We should thereby identify a certain *aesthetic of modelling* which, although it had a definite scientific background, was by no means entirely scientific. We might even become so abstract as to try to describe this aesthetic as the experience of a congruence or a consonance between the ways of the numbers in properly-functioning models and the observed ways of the waters of the physical world. It was essentially this experience that was shared between the members of the modelling group and the small band of connoisseurs that constituted its clientele.

In order to break out of the market restrictions which followed from the one-off, custom-built construction of numerical models, it was necessary to set up software 'production lines', for at least the 'batch production' of models. By these means, any model of a particular class could be constructed quickly and reliably using already-programmed procedures. These systems, which can be described as 'tools for building tools', were initially called 'Design Systems' (*e.g.* Abbott *et al,* 1973), but subsequently they became better known as 'Modelling Systems', or, to make the point more strongly, 'Industrialised Modelling Systems.' To the writer's best knowledge, this last denotation was first introduced by F. Holly (*e.g.* Cunge *et al,* 1980). They can also be described in the manner established independently in Artificial Intelligence, as 'modelling shells'. The institutional paradigm was in this case established by the Danish Hydraulic Institute (DHI), which in 1970 formed its 'Computational Hydraulics Centre' specifically in order to pursue this approach to modelling. This approach then constituted the third generation of modelling. It necessitated much greater investments in order to set up the 'software factories' in the first place, but thereafter models could be built at a fraction of the cost and within a fraction of the time required for the one-off, custom-built models of the second-generation approach.

The third-generation approach to modelling provided a number of benefits. It led to an increase in the number of clients for model results and a step-by-step development in the faculties of discernment of these clients. This permitted a considerable increase in the turnover of modelling businesses and thus provided the support for a larger number of individuals with a correspondingly greater range of backgrounds and talents. On the software side, it allowed the development of much more refined and efficient methods, which in turn justified the application of the system-built models to an increased range of practical applications. This is to say that the 'performance envelopes' of the systems were constantly improved. Moreover, the technological development was no longer confined within the minds of the individuals of the modelling groups, but it became 'fixed at the level of the code'.

The practice of third-generation modelling imposed further a certain unifying influence on the modelling group, making of it a true team, while it also permitted a standardisation of input and output facilities, data-base facilities and other auxiliary services. The paradigm was also one in which the modelling group carried through all the modelling work, from the first client contact through to the presentation and evaluation of the results, so that the modelling systems were constantly updated on the basis of practical experience. These systems then themselves 'learnt from the market', even as, in their development, they 'tracked the market'. The usual procedure was then to introduce updates as these were required by practice, with the systems being reconstructed as and when the number and variety of updates had made the code too unwieldy.

Third-generation modelling had a number of drawbacks, however, and these again arose primarily from the restricted range of its markets. The modelling group necessarily became increasingly esoteric in its methodologies, its jargon and its almost mystical 'feeling' for the systems and the models themselves. It became increasingly difficult to communicate with the 'uninitiated' in this field. So long as the market was only one for the *results* of models, this situation was supportable, but already in the late 1970s a demand arose to provide the modelling systems themselves, and this naturally led to serious communication difficulties. It was already known that the transfer of a one-off custom-built model led to difficulties enough (*e.g.* Cunge *et al,* 1980, pp. 391-405), but the transfer of a complete modelling system increased these difficulties by an order of magnitude. Only when the system was transferred to an environment already experienced in the thought world of computational hydraulics, and only after training had been provided over periods of many months, could these systems be brought to perform at all satisfactorily. These difficulties greatly restricted the dissemination of modelling systems and acted as a powerful brake on further applications of numerical modelling in this field. Modelling was still restricted to a small number of clients for results, even if this number had now risen to be of the order of a few hundreds. Some typical outputs of models of this generation are shown in Figures 5 to 7.

1.3.3   Fourth-generation modelling

When seen in retrospect, the first three generations of modelling may be seen primarily as means of preparing the way to the fourth generation. We can perhaps best see how this preparation has proceeded by considering the developments in the digital-machine systems upon which the models of these generations were run. The first and the second generation began their lives running upon stand-alone machines which, by present-day standards, were exceedingly slow and exceedingly expensive. These machines were the luxuries

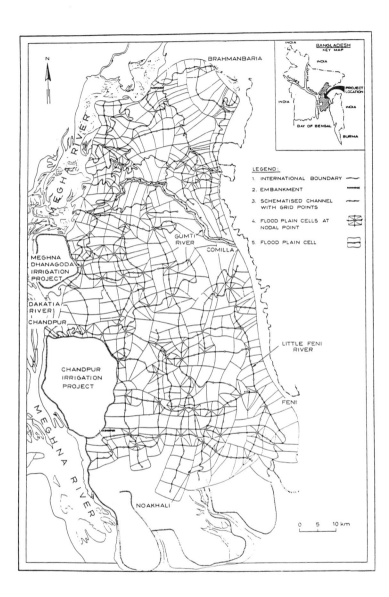

Note: River and flood plain schematisation for the South-East Region of Bangladesh. Models of this kind are now used for on-line forecasting purposes. In this case the root-mean-square errors of forecasts, over variations in water levels of many metres, are about 3.2 cm for 24-hour, 5.7 cm for 48-hour and 8.5 cm for 72-hour forecasts.

Figure 5   River and flood plain modelling

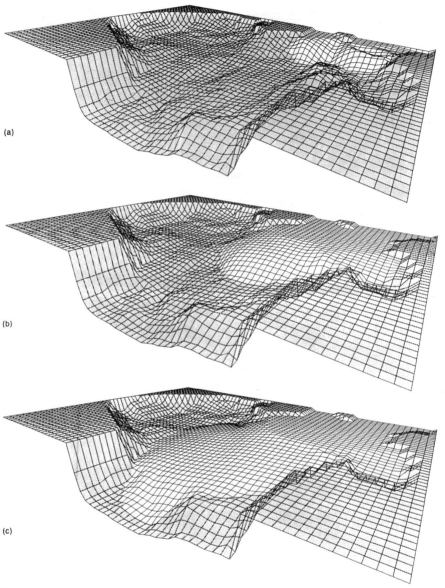

(a)

(b)

(c)

Note: Three 'snapshots' of the progress of a salt water front propagating over the Copenhagen-Malmö sill and down into the Baltic. The brackish surface layer has been removed for ease of interpretation and a vertical scale exaggeration is used.

Figure 6   Salt water intrusion into the Baltic through the Sound

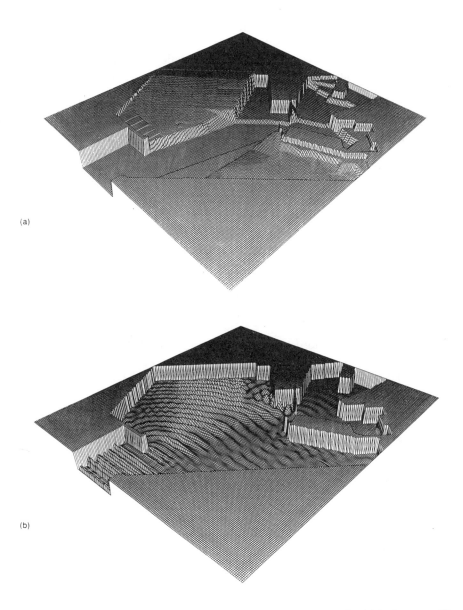

(a)

(b)

Note: The effect of constructions is modelled in the harbour of Valencia; (*a*) bathymetry; (*b*) a typical wave climate in the existing situation. Both illustrations are with vertical exaggeration.

Figure 7   Short-period waves in a harbour

of a relatively few modelling centres and their very cost restricted the dissemination of modelling outside of these centres. The third generation of modelling was based upon the widespread introduction of distributed computer systems, which began already in the mid-1960s with the advent of the IBM 360 series. By these means, a number of individuals could work separately with one common code and one common data base through common input-output facilities. The cost of the distributed computer systems still constrained the application of models, while restricting this development to individuals who were connected together by each such system. The distributed computer system thus served not only as a unifying element in the organisation, but as a mechanism for 'closing off' the group from its environing, 'average' world. For the present author's view of third-generation modelling, see further Abbott (1990(a) and (b)), and Abbott and Madsen (1990).

The development in information-distribution technology in our own times is often compared with that which occurred in energy-distribution technology during the last century. In this analogy, the stand-alone machine corresponds to the stand-alone steam engine, driving one pump, one steam hammer or one textile frame. The distributed computer system is then compared with the system of pulley wheels, belts and shafts by which energy was distributed in the factories of the last century. These developments now appear, however, only as preparatory to the introduction of the electric motor, together with the development of the electricity generating system and its distribution network. The widespread introduction of the electric motor is then compared with the widespread introduction of the microprocessor-based computer, or 'micro', with its various realisations, such as those of the Personal Computer (PC), or more latterly the Personal System (PS), and the Work Station. The network is then carried over also, becoming in the first instance a restricted Local Area Network (LAN). The massive introduction of the 'micro', and the reduction thus attained in cost per unit of computing and memory capacity, has been the prime moving force in the development of the fourth generation of modelling.

Fourth-generation modelling became a practical possibility with the introduction of IBM's PC/AT, with appropriate co-processors, in the early 1980s. This was the first machine to have the computing power to run a real-world simulation while being sufficiently standardised to justify the construction of a corresponding modelling tool. This tool could be directed to a much-extended market, but it then had to take account of the fact that this market was not composed of computational hydraulics experts. It thus had to provide facilities for setting up any given model of a certain class and for giving some help and even tuition during the setting up, it had to warn its users of possible set-up errors and, if the user could not provide certain parameters, it had to provide reasonable default values. Moreover, it had to be fault-tolerant, in that, if the user did make a mistake, this should not lead to the erasing of files and other losses of data. The tool that resulted could then be described as a modelling system with a complete menu-driven user-interface, together with some built-in advice servers and advanced graphics facilities allowing for extensive checking of both input and output. Extensive data-base facilities were also commonly incorporated. Examples of such tools are currently the WASP and WALRUS packages of Hydraulics Research in the U.K. and the MOUSE and MIKE 11 of DHI. Although applications of these systems are currently restricted to networks of one-dimensional flows, systems for simulating two-dimensional flows have already appeared, while it is anticipated that a whole range of such tools will be in use by the mid-1990s.

Although originally developed for the most common PCs, systems of this kind are now being increasingly mounted on more powerful Work Stations. The resulting increase in scope for their operation then makes them suitable for Computer-Aided Design (CAD) applications. It also, however, opens up perspectives for quite new uses of models, as will be outlined later.

One further aspect must be mentioned at this stage, however. This is the quite radical change that has occurred in the education of the clientele for modelling services, at least in some places and in some disciplines. Whereas the number of people with a real taste for quality modelling, world-wide, was only of the order of a few dozen during the 1960s, and had perhaps risen to the level of a few hundreds in the 1970s, it has now already risen to many thousands. This is not so much because of any special development at the level of formal education, but more because of a widespread awakening of interest in computers and computing and an overall increase in awareness of the capabilities of such tools in general. We have here to do more with a supportive *cultural* transformation. The fourth generation of modelling moves forward on the basis of this awakening of interest and this increase in understanding, so that the use of modelling spreads to scales of social organisation that it could previously never have reached. We have here to speak, in relative terms, of a 'mass market.'

There is a further difference between fourth-generation modelling and the previous generations. This is that whereas the first three generations were directed to providing modelling services by their builders through the performance of *projects*, the fourth generation is directed by its builders to providing services through the providing of *products*. The first three generations provided the modelling groups and the technologies with which to proceed with fourth-generation modelling, but this last form transforms the groups themselves from performers of projects to producers of products. This constitutes a very fundamental step which has all manner of institutional consequences — and certainly far too many for them all to be considered here.

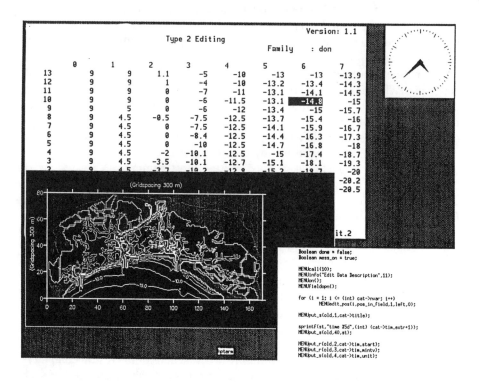

Note: This 'screen dump' from a modern fourth-generation simulation-modelling tool serves to illustrate the 1990 status of two-dimensional modelling. In the situation shown here (a coarse grid model of the Venice lagoon), the user has just updated some local bathymetric data, as held in one window, while holding a contour plot of the whole region. In this case a window has been opened by the system developer on a fragment of the source code, illustrating how such a system can also serve as a tool in its own development.

Figure 8   Screen view of a fourth-generation system in use

The advent of the fourth generation of modelling, however, marks the opening up of a new division within hydraulics between those whose primary objective is to make modelling products and those whose primary aim is to use these products (Cunge, 1989). We thus observe the appearance of that division into 'tool-makers' and 'tool-users' that is already a familiar feature in almost all parts of industry and agriculture.

Not only does the production of fourth-generation modelling tools ride on the back of third-generation modelling on the people side, but it also depends almost completely on earlier work on the software side. In effect, the procedures that enter the new products incorporate all the learning accumulated in the earlier forms of software. Without this transfer of experience at the level of the code, the cost and the development time of these new products would be prohibitive. Even on this basis, the costs involved in developing and marketing a sound fourth-generation modelling product are very considerable (*e.g.* well over a million dollars so far in the case of the MOUSE). It should be emphasised — even though it is probably obvious — that the product itself is of very little use unless it is properly supported by documentation, training and, most important of all, an on-line trouble-shooting service operating at the local level. An overview of fourth-generation modelling is given in Abbott *et al* (1991), while a typical screen view of such a system when being used for design purposes is shown in Figure 8.

1.3.4 The fifth generation

The fourth generation in modelling already constitutes a veritable revolution in hydraulics, in that the social benefits of hydraulics are thereby enormously extended and augmented. However, this revolution is really not itself a scientific revolution: the scientific revolution necessary at this stage had already been largely carried through in computational hydraulics against a background of second- and third-generation modelling. The fourth generation constitutes a revolution in another direction than the scientific, and indeed in a direction that follows more closely the social dimensions of hydraulics. Now it happens that the overall direction of this revolution was also mapped out long ago by 'the philosopher of technology' of our century, Martin Heidegger. Already, in 1927, Heidegger introduced the notion of 'the discourse of average intelligibility'. This has to do with the manner in which discourse really proceeds in human society, both at the general and the more specialised level. It leads into the study of how a certain body of knowledge can be transmitted within a world of discourse of average intelligibility. This is a world that, in the present case, knows little or nothing of the problems that wrack the computational-hydraulics experts; it does not understand their jargon and it is largely uncomprehending of their enthusiasms and their despairs. The problem of fifth-generation modelling is then that of expressing the knowledge and capabilities of the experts in terms of a discourse of another and more average intelligibility. In practice there will then be several such fields of average discourse, corresponding to different areas of interest and intent, each with its own set of 'common-sense truths'. The
ss of fifth-generation modelling has to do essentially *with raising the level*

of discourse of its clients, thereby increasing the competitiveness of these clients in their Heideggerian 'mundane and average world'.

The first means for solving this problem of fifth-generation modelling, and so of realising its aims, the tool which is most appropriate to it, has already been described and characterised as a product. This product then *encapsulates* the knowledge and capabilities of the computational-hydraulics experts in such a way that it can already enter into a discourse that, although of a more average intelligibility than that of the professional computational group, none the less imbues its owners with higher capabilities. However, it then quickly appears that the tool that we have so far described, which is essentially a simulation tool, is only one of the wide range of tools suited to this more general purpose.

A computer code is a sequence of signs, and Heidegger explained how each and every sign acts as 'tool for raising-up tools into activity'. The signs of the code raise the code up into activity as a tool, and this tool raises up other tools again, ultimately leading to such activities in our outer world of action as the construction of works and the implementation of operating policies. Indeed it is only at the very last stage that computers and computing contribute anything of material benefit to society. A vital level in this process is then that of *human decision-making*. The fifth generation of modelling has to do mainly with this level, of *human decision-making*, in the process of raising-up tools. Moreover, it has to do with decision-making at all levels in society. It thus has to come to terms with 'common-sense truths' in many areas of application and at several levels within society.

The fifth generation takes further the problem of conducting a discourse of average intelligibility by communicating in wide subsets of natural languages, alongside more conventional number descriptions, and it inseparably applies elements of intelligent behaviour during this operation. Thus fifth-generation modelling in this field can be characterised as a fusion of earlier work in the area of Computational Hydraulics (CH) and work in the area commonly referred to nowadays as Artificial Intelligence (AI):

$$\textit{Fifth-generation modelling} = (\text{CH}) \cup (\text{AI})$$

The potential products of fifth-generation modelling in hydraulics generally appear already to be endless, ranging from micro-climate control systems for public and private buildings and horticulture to integrated environmental management systems for the North Sea, the Mediterranean and the Arctic Ocean. (See, for example, Kowalik, 1986.)

If we now enquire into the role of modelling generally in this area, however, we find that it provides, in effect, the 'central interface' between domain data, such as is provided by point-monitoring stations, weather radars, satellite image processors and so on, and the human decision-maker. What was previously 'the numerical model' now becomes the *domain knowledge encapsulator*, for it is this which now encapsulates all that is known about the physical system that can be taken as given in any particular situation. This encapsulator must then itself be capable of using all data that are available and of processing them in such a manner and with such rapidity that they can be efficiently assimilated by the human decision-maker. We see that what was originally 'the numerical model' now appears as the nexus of a *hydroinformatics system.*

As presently conceived — for we now speak more of intentions, such as are by no means yet fully realised — these tools must make use of the 'enabling technology' that currently goes under the name of Expert Systems (ES) or, usually more appropriately, Knowledge-Based Systems (KBS). It must, however, make use of this technology in ways that are very different from those currently pursued in ES and KBS practice. At the same time, hydroinformatics takes over the main function of these tools, which is that of providing a marked improvement in the working performance of their users, and so of providing them with an increased productivity and thence enhanced competitiveness.

We should further observe that the way we have followed above is still only one way of approaching hydroinformatics. It is also possible to approach this subject from the side of data-base technology, in which the model appears only as a 'transfer device' within the data processing, or from the side of AI *per se*, where the 'model' appears even more as a means of encapsulating 'deep knowledge'. We shall return to these aspects later.

To summarise this description of the transformation of computational hydraulics into hydroinformatics, we may use the analogy of three mutually-orthogonal directions or axes of progress. The first three generations then provided primarily an advance along the axis of *depth of scientific development*, the fourth provides primarily an advance along the axis of *breadth of application*, while the fifth provides primarily an advance along the axis of *height of competitive potential*.

1.4   Towards the specification of a hydroinformatics system

The pragmatic task of hydroinformatics is to combine and integrate all standard-scientific information and, further, all common-sense truths to the extent that these truths can be given a standard-scientific informational representation. The combinations and integrations can, of course proceed in any number of ways, corresponding to the various fields of applications and levels obtaining within these fields. Let us, however, now list the main informational and common-

sense-truth features of a typical water-resources system that may have to be so
represented:

1.1   The international community and national laws and regulations that must
        be obeyed everywhere and at all times during the operation of the
        system.
1.2   The local bye-laws that are effective in specific areas at all times.
1.3   Laws and bye-laws of a temporary nature.
1.4   Other local and temporary constraints of a law-like or regulation-like
        character (*e.g.* institutionalised policy, contractual agreements, local
        precedents, established social norms and mores).
1.5   The effectively-constant physical, chemical and biological parameters of
        the specific area under consideration.
1.6   The meteorologically-determined, and so often rapidly-varying flows of
        water, sediments, chemicals and other water-borne substances in the
        geographical area of concern.
1.7   Measurements of flows, transports and water-quality indicators available
        within the basin.
1.8   The sites and quality parameters of all water users in the area.
1.9   The uptake and storage-basin characteristics of these.
1.10   The sites and production rates of heat, chemical and biological polluters
        in the area.
1.11   The discharge potentials and retention-basin capacities of these.
1.12   The states of all control elements (control structures, pumps, retention
        basins, treatment plants generally) within the local area, and the
        relations between them.
1.13   The 'initial-condition' distributions of heat, chemical and biological
        indicators in the area.
1.14   Other qualified and quantified aims and objectives of the system.

An initial feature of such a hydroinformatics system is that it allows the
use of numerical simulation subject to constraints that are expressed in natural
language — such as are all forms of legislation, contracts, agreements, etc.
When seen as a knowledge-based system, the hydroinformatic system provides
the usual possibilities for its user to pose questions, such as 'how?' and 'why?'
and 'when?' For example, when confronted with the information that he must
reduce discharges of his plant to a local watercourse of a certain waste product,
and so store or otherwise dispose of the remaining part of the discharge, the
manager of the plant may naturally ask: 'Why?' He is then provided with a
word description — generated by using existing KBS procedures — of the
reasoning, the measured data and the numerical simulation results that have led
to this decision. For example:

Because
The law of 15 October 1981, section 4, paragraph 3 limits concentrations of aa in natural streams to bb mg/l.

and
The weir at A is partially closed for repair until 5 March 1989.

so that
Flow is limited in the B river.

and
The cc concentration meter at C already gives a reading of dd mg /l.

and
A calculation shows that the limit of concentrations will be exceeded at D unless discharge is reduced to ee mg/s.

Following on from this, we may now enumerate the principal characteristics of the products of hydroinformatics in such a case, which we have already introduced under the rubric of 'hydroinformatics systems':

2.1   They commonly constitute combined knowledge-based and algorithmic-based systems for both the off-line and the on-line design, control and management of aquatic environments.

2.2   The principal pragmatic goal of hydroinformatics is to reduce the total costs to society and to ecosystems generally of industrial and agricultural operations in so far as these operations interact with the aquatic environment. Thus, in many cases, hydroinformatics systems will incorporate optimisation procedures.

2.3   The most common form of the hydroinformatics product, to date, is the expert-advice system. In this case it is not normally acceptable that the system should itself make decisions about the operation of a water body and its environment, so that it does not normally have to overrule any human decision.

2.4   General hydroinformatics tools will be designed in future with the flexibility to answer to different concerns. These concerns will be in principle of equal importance, but each may in a given situation be the dominant concern.

2.5   It is observed already that measuring and operating systems become as far as possible interfaced electronically to the domain knowledge encapsulator and its frame. Similarly, the complete hydroinformatics system is accessible. There is then a two-way communication: monitoring systems may generate signals (which may be interpreted as alarms) to the system and the system may request information from subsystems as well as from its information and knowledge bases.

2.6   The main users of such systems are those charged with the management of the aquatic environment. It is possible that other users may be admitted, such as those more directly responsible for stressing this environment.

2.7   It is expected that, in the immediate future at least, its users will continue to access such systems through computer terminals.

2.8   These systems are activated for the most part through the execution of commands leading to prioritised interrupts from the terminal. They may however also be event-driven. Real-time events interact with these systems and they will themselves respond using a hierarchical system of priorities in each application area. The user may enter additional information when this is called for or he may invoke built-in system functions, either on his own initiative or on the suggestion of the system itself.

2.9   Learning and updating facilities have still largely to be incorporated. Initially, updating will almost always occur off-line.

2.10   As remarked earlier, interrogation facilities will be provided.

When such a functional specification is reviewed it is naturally costed. What then emerges is the very great cost of such systems, which is in its turn a function of their immense complexity. Of course, this cost will be recouped in time by the benefits that these systems confer, but still the initial investment is already in many cases substantial. A review of the workloads involved in other 'high-technology' areas, such as are available in the case of the European Commission's ESPRIT projects, provides estimates of future investments of between 20 and 500 man years in a single 'working environment'. Such quantifications indicate that the full development of hydroinformatics will necessitate the formation of much larger groups and the accumulation of a much wider spectrum of expertise than any that have so far been assembled in this area for modelling purposes. It also indicates that another way of working will be necessary, as compared with that so far practised in numerical modelling. In particular, it is clear that hydroinformatics systems will have to be built up from proprietary codes that have been constructed, in most cases, for quite other purposes than those of hydroinformatics (*e.g.* Abbott *et al*, 1988).

Looking at this matter more generally, or from a greater distance, we observe that man's ability to intervene in the affairs of nature has been growing at a faster rate than has his ability to manage these interventions, and this for some centuries now. The historic task of hydroinformatics is then to redress the balance in this respect, and this can scarcely be a simple or a cheap matter at this stage.

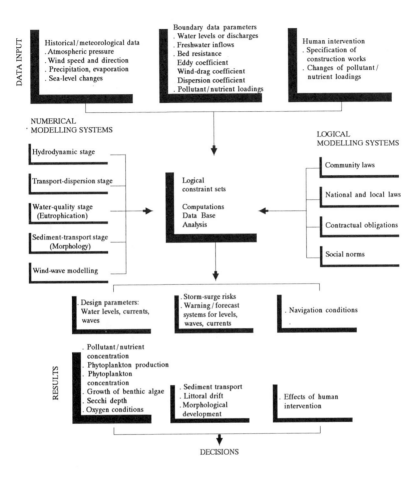

Note: Schematisation of the overall structure of a late 1980s prototype hydroinformatics system. It will be clear that the buffer memory for the routine surveillance data stream needs to have properties different from that which is storing legislative constraints and other natural-language-expressed rules (in this case, a structured-query-language (SQL)-type data base as opposed to an object-orientated data base: see, for example, Date, 1989). Similarly the natural-language-expressed facts and rules, with their merging, inference and interrogation facilities, have to be organised using KBS shells. In general, hydroinformatics systems must integrate a number of different languages, environments and tools — most of which have not been designed specifically for hydroinformatics applications — into particular and usually custom-built environments.

Figure 9 A prototype of a relatively complete hydroinformatics system

The evolution of this specification is best followed here in terms of illustrated examples. The coupling of measuring instruments to a buffer data base, and thence to a single SQL data base where this data can be further processed and analysed, is illustrated in Figure 10. Systems of this kind were realised already in the early 1980s and prepared the way for the introduction of integrated monitoring, data modelling and process modelling systems, as will now be illustrated here in some detail for the coupled hydrographic monitoring and forecasting system realised towards the end of the 1980s for the Great Belt rail- and road-link project. At an estimated cost of about 3.7 billion dollars, this is currently, after the Channel Tunnel, the largest construction project currently proceeding in Europe. As the Great Belt carries almost all exchanges of waters and water-transported materials as well as almost all sea traffic between the Baltic and the North Sea, interventions of this kind should have, even in the worst case, a neutral impact on these exchanges: this requirement is in fact embodied in the corresponding enabling legislation, as approved by the Danish Parliament in 1987. Because of its position between the fully-saline North Sea and the much-less-saline Baltic, the waters are commonly quite highly stratified, with lighter, brackish water layers situated on top of heavier, saline water layers. Although the purpose of the Great Belt information system is mainly one of hydrographic forecasting, a certain amount of environmental control can be exerted through a more informed planning of dredging-spoil disposal and other such operations. We can thus regard this example as a prototype of a hydroinformatic system, even if still an incomplete one, and so it is worthwhile to describe it in more detail, following Kirkegaard et al (1991). In this way we can also make the otherwise rather abstract notions introduced in this book more concrete; we can, so to speak, bring them down to earth again.

The Great Belt hydrographic information system itself integrates component monitoring and forecasting systems. Its main components are thus an environmental monitoring system and a forecast modelling system. Data from the monitoring system are transmitted on-line to a central data base installed at the responsible hydraulic institute (DHI), some 125 km from the project site. The hydrodynamic forecast models, developed by DHI, are installed on a 'super computer' at the Danish Meteorological Institute (DMetI), which is also linked to the DHI data base so that the forecast model results can be updated with the most recent observations. From the central data base, the users can obtain information from a large number of sensors, as well as model-predicted values of currents and water levels.

The information can be obtained from a work station, linked to the data base, with menu-driven windows for plotting and other purposes, or from PCs coupled to the central data base by telephone modems. In addition, the daily hydrographic forecast is issued on telefax. This arrangement is schematized in Figure 11.

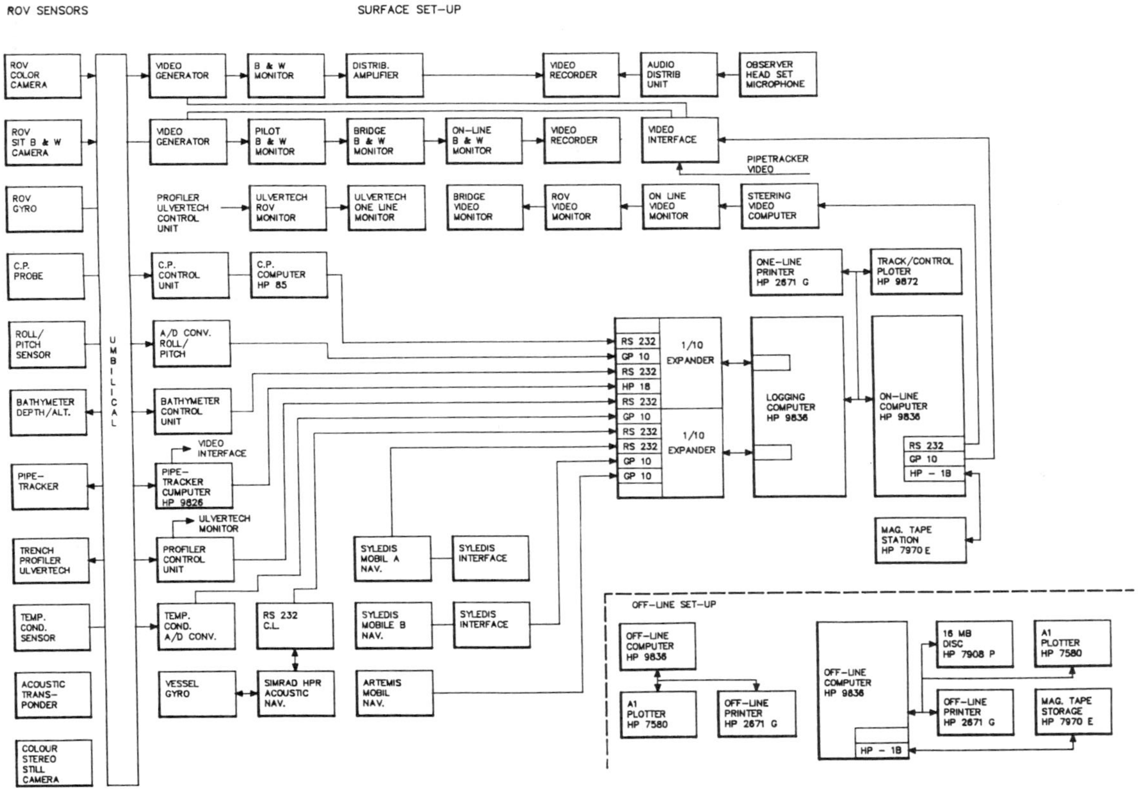

Figure 10 An early 1980s realisation of the measurement side of a hydroinformatics system: an instrument set up for a remote-operated vehicle (ROV) inspection

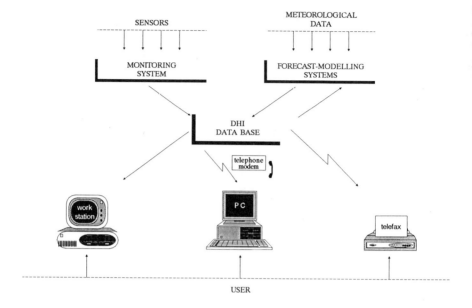

Note: The main components of the on-line hydrographic system being used during the construction of the Great Belt road and rail connection. Information is accessed, using menu-driven windows, either via a work station or through PCs. The telefax is used to record the daily hydrographic forecast.

Figure 11   System overview

The Monitoring System is used to provide data for the verification of numerical models, for a Vessel Traffic Service System (VIS), for monitoring of the environment, and for design and forecast purposes. (See Møller, 1989.)

The hydrographic and meteorological forecasts are delivered through this system so that it provides not only the modeller with data, but also the supervising engineer and the contractor with information for planning purposes.

The Monitoring System has been in operation since the middle of 1989 and will continue to function, with further developments, until 1996. It is itself comprised of eighteen permanent measuring stations and a ship-based station, as schematised in Figure 12. Such a monitoring programme necessitates the long-term deployment of instruments in the Great Belt: these instruments provide data on stratification (by measuring temperature), salinity, oxygen and turbidity over verticals. Vertical current profiles are obtained by Acoustic-Doppler Current Profilers (ADCPs) mounted both on the sea bed and on board the permanently-deployed survey vessel. An instrument for detecting the interface between the two water layers, the AIR (Acoustic Interface Recorder), was developed by DHI especially for this project. The AIR is deployed on board the survey vessel and at several sea-bed positions.

Water levels are measured at several shore stations located in the Great Belt region, north and south of the project site and at two light houses located at sea. Wave conditions are measured by wave riders located in two positions SW and NE of an island in the Belt (called Sprogø). Meteorological parameters (wind speed, direction, and gust at 10 m and 70 m altitude, air pressure and precipitation) are measured on Sprogø. Data are transmitted to DMetI, DHI and the Danish National Research Center (RISØ).

Figure 12   One of the permanent environmental monitoring stations in the Great Belt

The survey vessel performs a variety of tasks. These include the servicing and maintaining of fixed monitoring stations, the temporary replacement of malfunctioning sea stations, determinations of horizontal distributions of hydrographic parameters for the verification of mathematical modelling system results, and the tracing of sediment plumes during dredging operations. For these tasks it is equipped with a variety of profiling instruments and a suitable radio and data-communication link.

In order to pilot the survey vessel during plume tracings, custom-built signal processing software is interfaced with the ADCP so as to provide vertical concentration profiles of suspended sediments along the surveyed lines. Further to this, an on-line Lagrangarian sediment-plume model is installed. This can be run interactively in a 'now-cast' mode using the instantaneous current measurements from one of the fixed stations, or in a forecast mode using the prognostic current. This facility is used to plan near-field suspension and sediment monitoring.

Data collection and transmission from the various on-line sensors to the DHI data base is managed and controlled by a 'central control computer', which is a standard UNIX-based work station. This central control computer performs the following tasks at regular intervals:

o   Collecting data from shore-based systems.
o   Collecting data from the Sprogø data centre.
o   Providing timing information to all subsystems (GMT).
o   Preprocessing and organising of data.
o   Issuing automatic alarms for missing or corrupted data.
o   Providing data for on-line display to external users.

Check procedures are implemented in the system which, among other things, monitor the format of the data telegrams and the sensor-status parameters indicating instrument performance and ensure that data are within realistic intervals. In the event of any malfunctions being observed, alarms and warnings are automatically produced, displayed and stored in error and information logs. This system information is later used by the service team for correcting errors and for the planning of station service visits. A more complete description of these aspects is given in DHI (1990).

The forecast data available in the hydrographic information system are water levels and the current velocities in the upper layer of the Belt for a five-day period. The dynamic wind and air-pressure distributions combine with the tidal motions to form the hydrographic conditions in the Great Belt; but also the density differences between the brackish Baltic Sea and the southern North Sea water contribute their influences. By the use of coupled meteorological and hydrodynamic models and extensive use of the monitoring system, it has proved possible to predict the water levels and currents for these complex hydrographic conditions with adequate accuracy.

The numerical weather prediction models originally used were the UKLAM model (United Kingdom Limited Area Model), producing short-range forecasts (0-36 hours ahead), and the global ECMWF (European Centre for Medium Range Weather Forecast), from which forecasts for between 36 and 120 hours are available. The UKLAM was replaced in 1990 by the Danish Meteorological Institute's version of the HIRLAM model (High Resolution Limited Area Model), describing the meteorological situation over Denmark in more detail. These models provide wind speed and direction (10 m height) which, together with the air pressure, are used as input to a third-generation hydrodynamic modelling system (SYSTEM 21, or S21), already developed by DHI, describing depth-integrated two-dimensional flows. Three computational domains are set up: a coarse grid domain (grid size 18520 m) covering the North Sea and the Baltic, a less coarse grid domain (6173 m) covering the transition area, and a fine grid domain (2057 m) covering the Danish Belts. In this total model complex, the hydrodynamic equations are solved simultaneously in the three computational grids. The coarse grid is used to pick up the oceanic-charted boundary conditions, which are specified by the tidal level adjusted with the oceanic atmospheric pressure forcing terms. The wind shear stress and atmospheric pressure fields constitute the internal, free surface boundary condition.

The output from the hydrodynamic model is composed of predicted water levels and currents averaged over the water depth. These outputs are then compared with measurements made at four locations in the Great Belt, in two at which water levels are measured and, in the remaining two, currents.

Experience has shown that the variation in time of the mean flow in the Great Belt follows the variation in the upper layer, but can be different in strength. A correlation procedure between measured and calculated time series has therefore been established. The procedure consists of a simple linear regression correlation between measured and hindcast model data from the preceding day. The result from the regression is then used to calibrate the forecast. In addition, as an alternative, a Kalman filtering technique is used to assimilate measurements and model results to forecast current velocities. The application is based upon Harrison and Stevens (1976), considerably augmented by Heemink (1986; see also Heemink and Kloosterhuis, 1990). For a detailed description of the forecast system, see Vested et al (1990). It should be added that for situations in which density variations in the vertical are important, use is made of a two-dimensional, two-layer modelling system (S22), some typical output from which has already been illustrated in Figure 6.

The user interface installed on the central computer is a generalised data display and management package developed for use on engineering work stations — the X-Windows Display System (X-DISP). The package is based on common industry standards for graphical user interfaces, and enables the user to build a custom-built, interactive, graphical user interface as a from end to any on-line monitoring or similar system. The following features are included in this custom-built application:

o   A user-configurable base map including a zoom facility whereby areas of particular interest can be provided to larger scale and in greater detail.
o   A user-configurable sensor station location overlay.
o   A user-configurable sensor station definition.
o   A library of plot routines for common data set types (time series, profiles, etc.)
o   An interactive graphical editor for time series data.
o   An interface to a SQL data base system for data storage.

X-DISP has been developed specifically as a generalised package for building custom-built user interfaces for hydrographic or hydrological monitoring systems. However, the system is easily adapted for use in connection with any other kind of on-line monitoring system. It is by design an 'open' package, allowing the user to add new elements, such as new sensor types, custom-built plotting routines, and new communication modules.

Apart from this more central use of the system, external users can, with the help of a PC, gain access to the data base as follows:

1.   The user starts a program on the PC: the program displays a screen picture providing access to a number of menus, which enable the user to specify the data required (stations, data types and time period).
2.   When these have been specified, the program automatically calls a DHI host computer, and specifies the necessary user name and access codes to establish access to the system.
3.   The selected data are then automatically transferred back to the user's PC, and the telephone link is closed.
4.   The user can now choose to make use of the built-in presentation programmes, to show the selected data on the screen, or to process the data further according to requirements. The built-in programs can display the received data, primarily as plots, or generate a print-out.

In the almost purely forecasting mode of operation of this last example, numerical modelling systems are combined with a monitoring system so that forecasts can be updated or, more generally, so that measured data can be *assimilated* into the total forecasting process. We are in such a case, however, only looking into a future, and indeed predicting a future, over which we have ourselves little or no control. There are other situations, on the other hand, in which we may ourselves have a large measure of control over the future, and we can then cause this future to correspond rather well to our best-perceived interests. This last situation occurs in fully-regulated channel systems and, to some degree at least, in many storm-water sewer systems. Between these extremes we have a whole range of situations in which we may exert some influence — we may 'nudge' the physical system in one or the other direction — but we cannot completely control it, so that we are somewhere between a 'pure' forecasting situation and a 'pure' control situation. We then observe that as we move through this range the nature of data assimilation changes markedly. Thus, in the pure forecasting mode, we commonly assimilate monitored data by merging numbers that we have computed with comparable numbers that we have measured, using weightings that reflect our estimates of the relative reliabilities of these numbers. Clearly this procedure is no substitute, as and by itself, for poor process modelling and, as we pass over to control situations, it needs to be augmented with other methods. The difference as we proceed in this way is schematised for the case of hydrologic modelling in Figure 13.

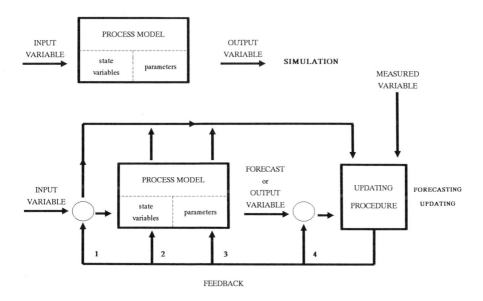

Note: An organogram illustrating the development from a stand-alone simulation system to a forecasting system in which simulated results are modified on the basis of on-going measurements.

Figure 13   Updating techniques in hydrological modelling

In many cases discrepancies between computed and measured values cannot be reconciled at all in this way since these discrepancies can only be accounted for by some major structural change in the behaviour of our physical system. For example, a dyke may break, causing extensive flooding, or a sewer pipe may block, or a pump break down completely, or whatever. In such cases we can best proceed to a knowledge-based updating, in which a knowledge-based system (realised using a standard KBS shell) intercedes knowledge between a data base and a numerical modelling system. A schematisation is given in Figure 14.

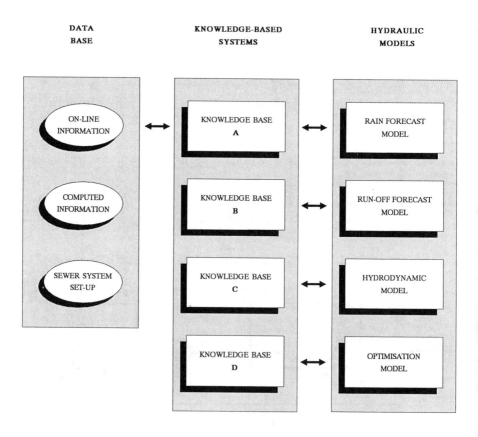

Note: Overview of a combined data and knowledge-based forecast-updating and optimisation system.

Figure 14   System overview

As an example of knowledge-based updating we may consider the operation of diagnostic systems for hydrodynamic flow networks (Amdisen, 1991). We have a hydrodynamic flow network upon which is superimposed an information flow network. We can describe both networks in terms of objects, while we can define the states of connectivity between the nodes of the hydrodynamic flow network in terms of Boolean arrays. In practice, use is made of some elements of classical graph theory by introducing a *successor*, or *adjacency matrix*, which defines which nodes follow a given node, and a *reachability matrix*, which defines which nodes can be accessed in any way at all from a given node. (More generally, the connectivity properties of complex nets of the kind already introduced here in Figure 1 are described in terms of topological invariants of the net — such as its Betti numbers, and these in turn define algebraic invariants, such as the ranks of the node-residual matrices used in the numerical solution algorithms: see Abbott, 1979, p. 216; Itai and Roden, 1984; Ormsbee and Kessler, 1990).

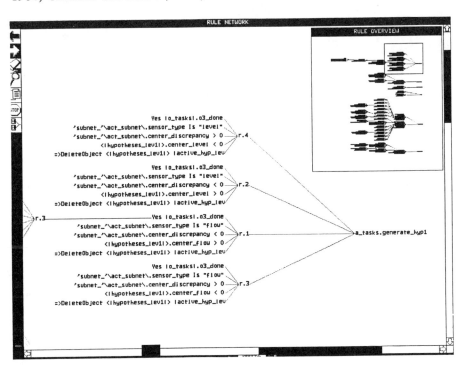

Note: The illustration shows two windows onto the inference net used for on-line diagnostic analysis of a storm-sewer system.

Figure 15   Application of a KBS to SSS fault diagnosis
(continued on next page)

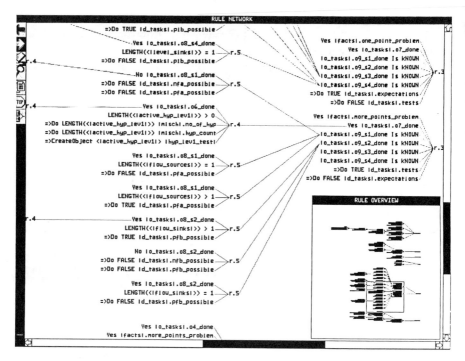

Figure 15   (continued from previous page)

When a marked discrepancy is observed between computed and measured results, a subnetwork that includes the discrepancy is extracted. The node nearest to the point of discrepancy is chosen as a 'centre'. Then the set of all hypotheses that are consistent with the observed discrepancy is activated, in practice by deactivating all non-consistent hypotheses. The points which could influence or be influenced from the chosen centre are identified through the use of the successor and reachability matrices. This causes us to begin monitoring variations between computed and measured results at points which could be influenced, and with definite expectations of when and how the variations should appear. The appearance of one such variation then sets the KBS to chain further so as to reduce the set of admissible hypotheses still further. Thus, as the process continues, we can narrow down the position and most likely cause of a structural change in our hydrodynamic system. The extention of this approach to comprehend other features, such as outputs of pollution indicators ('biosensors'), and to other applications, such as to the activation of control and other safety measures, is an on-going activity. Figure 15 illustrates one stage in the diagnostic process.

All of the above facilitaties are incorporated in the production on-line control system illustrated by the organogram shown in Figure 16. The first of these systems was being installed in the Danish city of Aalborg in 1991. For related developments in other fields, such as oilfield analysis and legal decision-making, reference may be made to Steele (1991).

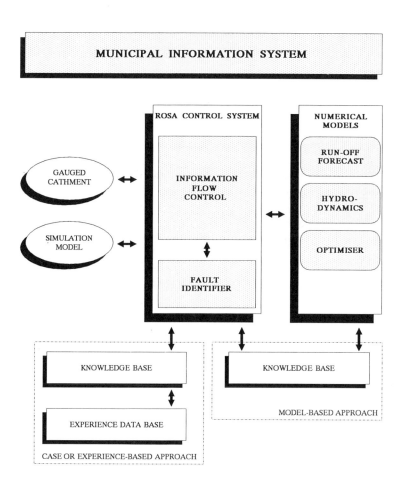

Note: Outline of a process control system that integrates an on-line data stream with numerical and logical modelling as a means of optimising the flow behaviour of a storm-sewer system.

Figure 16   The hydroinformatics system *anno* 1991

1.5   The commitment to abstraction

Hydroinformatics systems are usually, by their very nature, complicated. It is clear from experiences with the larger and more complex third-generation modelling systems alone that it will not be practical to construct hydroinformatics systems using the same organisational structures, both of people and of software, that have so far sufficed for numerical modelling. Indeed, third-generation systems like the *Système Hydrologique Européen*, SHE, already represent the outermost limits of the possible in terms of a unified computational-hydraulics group working exclusively in procedural languages. Problems of lead time, system reliability, system flexibility and development potential all indicate clearly the need for an approach based upon the attainment of higher levels of abstraction. The need for such an approach to complex systems was set out already in 1972, by Dijkstra, as follows (p. 122):

> We all know that the only mental tool by means of which a very finite piece of reasoning can cover a myriad of cases is called 'abstraction'; as a result the effective exploitation of his powers of abstraction must be regarded as one of the most vital activities of a competent programmer.

> ... The purpose of abstraction is not to be vague, but to create a new semantic level in which one can be absolutely precise. Of course I have tried to find a fundamental cause that would prevent our abstraction mechanisms from being sufficiently effective. But no matter how hard I tried, I did not find such a cause. As a result I tend to the assumption — up till now not disproved by experience — that by suitable application of our powers of abstraction, the intellectual effort required to conceive or to understand a program need not grow more than proportional to program length... The by-product was the identification of a number of patterns of abstraction that play a vital role in the whole process of composing programs.

Viewing this same matter from the point of view of data bases, Hoare (see Dahl *et al,* 1972) expressed himself similarly (p. 83):

> In the development of our understanding of complex phenomena, the most powerful tool available to the human intellect is abstraction.

By abstraction we create a hierarchy of codes, with any one level in the hierarchy working on and through other, lower levels. In the words of Dijkstra (1972, p. 125):

> Hierarchical systems seem to have the property that something considered as an undivided entity on one level is considered as a composite object on the next lower level of greater detail; as a result the natural grain of space or time that is applicable at each level decreases by an order of magnitude when we shift our attention from one level to the next lower one. — As a result the number of levels that can be distinguished meaningfully in a hierarchical system is kind of proportional to the logarithm of the ratio between the largest and the smallest grain, and therefore, unless this ratio is very large, we cannot expect many levels. In computer programming our basic building block has an associated time grain of less than a microsecond, but our program may take hours of computation time. I do not know of any other technology covering a ratio of 10^{10} or more: the computer, by virtue of its fantastic speed, seems to be the first to provide us with an environment where highly hierarchical artefacts are both possible and necessary. This challenge, viz. the confrontation with the programming task, is so unique that this novel experience can teach us a lot about ourselves. It should deepen our understanding of the processes of design and creation; it should give us better control over the task of organising our thoughts. If it did not do so, to my taste we should not deserve the computer at all.

In hydroinformatics, as in computer science generally, we can follow the increasing commitment to hierarchical systems in three major areas of computer science, namely, artificial intelligence, programming languages, and data bases. In all three cases we observe the development of the notion of *high-level modelling*. Thus, for example, Shaw (1980, p. 189) explained in general terms:

> What we call 'abstraction' in programming systems corresponds closely to what is called 'modelling' in many other fields. It shares many of the same problems: deciding which characteristics of the system are important, what variability (*i.e.* parameters) should be included, which descriptive formulation to use, how the model can be validated, and so on.

It then follows that, in hydroinformatics, we shall have to understand 'modelling' in this wider sense, of a general process of abstraction.

Within the three 'source areas' of hydroinformatics we observe, keeping to the same order as above, that knowledge representation, data abstraction and semantic data-base models all have common interests in high-level modelling, and then specifically for the construction of knowledge bases, for the elaboration of program and data-type specifications, and for the design of specific data bases. We observe, in computer science generally, the same convergence of interest around a specific core of knowledge and understanding that we have already observed in hydroinformatics.

Morgado (1986) points this out at the very beginning of his monograph (p. 2) as follows:

> Researchers in each of these fields have been using ideas from the other two fields in their work. LISP (the language most used in AI) has strongly influenced programming language research (Liskov, 1974) and AI concepts used in knowledge presentation (*e.g.* large irregular heterogeneous data structures and data sharing) and in implicit control of inference have influenced both programming language and data base research (Baroody and de Witt, 1981). Programming language concepts used in high-level languages, data abstraction and formal specifications are being used in AI and data bases (Lockman *et al*, 1979; Lowenthal, 1971; Melkanoff and Zamfir, 1978; Smith and Smith, 1977). Similarly, concepts such as type hierarchies, multiple-typed objects, persistent data, data base models and data base constraints, which are used extensively in data bases, are also being used in AI and programming languages (ACM, 1977, 1981).[3]

When seen from a general, computer-science point of view, (*e.g.* Cooke and Bez, 1984), this three-fold approach to abstraction is seen to correspond, within the hydroinformatics context, to the three basic elements of computer programming, namely, operations, data, and control. It is then only that in numerical-simulation modelling we put the emphasis on operations, and increasingly on operation abstraction (primarily, so far, through the elaboration

---

[3]The references given here by Morgado are listed among the references given at the end of this book. For an excellent glossary of common terms in AI, reference is made to Harmon and King, 1985, which work indeed provides a quite outstandingly clear introduction to Expert System (ES), or Knowledge-Based System (KBS), technology generally.

of hierarchies of procedures) while in data-base technologies the emphasis is on data abstraction and in AI (and especially in LISP) on control abstraction.

Further to this, we observe that the main axis of development in second- and third-generation modelling is on the side of the operations — albeit with control aspects developing in third-generation modelling to such a degree as to demarcate third-generation from second-generation modelling.

In fourth-generation modelling, the operation side advances rather little, but the control side takes a major step forward, if only because of the interaction with the menu system, and the data-base aspects also take on an own significance. None the less, the computational-operational part and the control are still maintained within one or at most two languages (FORTRAN primarily for the computational part, C primarily for the menu-interactive control part in MIKE 21: see DHI, 1990) and only one class of data base is employed. In fifth-generation modelling, the control and data-base aspects move decisively in the direction of abstraction, with more than one language, environment and tool, and different kinds of data base, integrated into hydroinformatic systems.

It is then clear that the entities and interests that are coming together in hydroinformatics are also coming together in computer science. When hydroinformatics is regarded in the restrictive, Husserlian manner, as an 'applied', or *normative science*, then computer science can be regarded as the corresponding *theoretical science*. The different approaches to the nexus of the theoretical science introduced here then have their own individual correspondences in the normative science.

The commitment to abstraction has the consequence that hydro-informatics must be founded, on this side, at the greatest possible depth of abstraction available in modern science. This is to say that it must be founded upon Mathematical Logic and Set Theory — and the relation between these (Manin, 1977; Genesereth and Nilsson, 1987; Thayse, 1988). Further to this, we observe more generally (Abbott and Basco, 1989) that computational hydraulics, when taken in the direction of computational fluid dynamics, is itself increasingly founded on the bases of Mathematical Logic and Set Theory.

We should then observe at this point that by far the best-known of abstraction methodologies currently used in computer science is that of *object-orientation*. In the earlier, and still most familiar methodology, which we may in retrospect call *function-orientation*, we think primarily in terms of actions or tasks, thereby relegating entities and events to a secondary role, and indeed the most common programming languages bias our thinking in just this direction. In these languages we can have any number of variables, but we can have only a few variable types (namely, in FORTRAN, integer, real, logical, complex, and array). As Woodfield (see De Coursey, 1990) expressed the matter: 'Programs written under this constraint are similar to English papers written using any number of verbs and pronouns but only four or five nouns to represent all conceptual entities.'

In the object-orientated (in American English 'object-oriented') approach, we try to carry over to our codes the essence of the abstraction process that we carry out in our own minds. We accordingly create a software world of *objects* which mirrors the world that we ourselves experience. Just as in everyday life, then, we abstract from a particular instance of an object (*e.g.* 'the weir at Geesthacht') to the set of all such objects (in this case, 'weirs'). We call this more inclusive set the *object class*, so that we can identify any particular object as a particular instance of its object class by prescribing value data (in the above example, that it is at Geesthacht, to which we could add the physical characteristics of the weir also) and behaviour data (what this particular weir does or can be made to do, out of the set of all possible things that can be done with weirs). Then, to pursue this example further, just as a particular weir belongs to the object class of all weirs, so this object class can in its turn be considered as an object in the object class of all control structures, and so on recursively, in principle. Value and behaviour data that are carried over from an object class to a particular instance of that class are then said to be *inherited*.

In the same vein, we can augment our world of hydraulic objects with a particular control system for a particular weir, which again would belong to the object class of all weir control systems, and so on again. The point is, however, that by proceeding in this way we are able to write descriptions that can be made complete, precise and unambiguous, and this can be done independently of any particular implementation of the description in code. Indeed, to follow this last point, we can write our descriptions in any language we like that balances the need for satisfying formal, consistent conditions of syntax against semantic convenience, *i.e.* ease of interpretation in our outer world. Having done this, we can manipulate our objects without any knowledge of their internal arrangements, so abstracting ourselves from their internal organisation. Thus, to keep to our example, we could request 'reduce flow over the weir at Geesthacht to $5m^3 \ s^{-1}$', and this would be done without our needing to know further how it was realised in the code.

We call such an abstract, formal description or specification of a set of values and behaviours an *abstract data type*. In his excellent *Object-Oriented Software Construction* (1988, pp. 53, 54), Meyer introduces this concept as follows:

> A data structure is thus viewed as a set of services offered to the outside world. Using abstract data type descriptions, we do not care (we refuse to care) about what a data structure *is*; what matters is what it *has* — what it has to offer to other software elements. This is a utilitarian view, but the only one consistent

with the constraints of large scale software development: to preserve each module's integrity in an environment of constant change, every system component must mind its own business. It must only access others' data structures on the basis of their advertised properties, not the implementation that may have been chosen at a certain point of system evolution.

... In summary, an abstract data type is a class of data structures described by an external view: available services and properties of these services.

At the present stage of development, which is mainly concerned with the design, construction and marketing of fourth-generation systems, both function-orientated and object-orientated approaches are employed, the first primarily for the computational components and the second primarily for the control components of the corresponding modelling systems. Similarly, object-orientation already plays an important part in the formulation of data types, and indeed in data typing generally. Alongside the use of inheritence mechanisms, it may further be advantageous to link together the calling of an operation and an object to which this operation is to be applied during the course of the running of the system, and in this case we have also to introduce a mechanism of *message passing*. Message-passing architectures appear to be necessary already for fifth-generation modelling. The further development of hydroinformatics, however, requires much more far-reaching reformulations again, such as will be introduced later in this book.

2 The social dimensions of hydroinformatics

Well one knows that pure thirst is only quenched in pure water. There is something exact and satisfactory in this matching of the real desire of the organism with the element of its origin. To thirst is to lack a part of oneself, and thus to dwindle into another. Then one must make good that lack, complete oneself again, by repairing to what all life demands.

Paul Valéry, *The Sources (Les Sources)*

2.1 Hydroinformatics and the natural environment

2.1.1 Our more basic motives for modelling and controlling the natural environment: the social need for hydroinformatics

The issue of the protection of the natural environment is today one of the most pressing of social and political issues. It is now clear that the progress of industrialisation (including the industrialisation of agriculture) has been purchased to a considerable extent at the cost of the natural environment, and indeed to such an extent that fears for the future of this environment increasingly act as primary constraints to further industrial and agricultural development. Within the European Community, for example, attention is currently being given to a wide spectrum of environmental problems, and several Community-wide projects have now been initiated for studying these problems. The central theme of these studies is that such problems cannot be solved without an understanding of the physical, biochemical, ecological and human-social processes that influence our natural environment. A particularly important emphasis is then placed upon the marine and aquatic environment, and thereby upon hydroinformatics.

When we first face these environmental problems, when we first confront them in their full natural and social-economic complexity, then we cannot but feel overwhelmed by their magnitude when seen within the context of a modern technology. The massive destruction of natural species, the terrible suffering which we recognise in the larger mammals, such as the seals, the dismal desertification of the sea bed, the poisoning of the soils and all manner of other manifestations besides, cannot but induce a state of deep consternation, mounting up to anguish, in any individual who becomes truly aware of the nature and extent of this environmental catastrophe. With this consternation, with this anguish, we experience a deep sense of guilt for this disaster that we have brought down upon ourselves and our fellow creatures. In particular, we see in this the ultimate dereliction of our stewardship over our fellow creatures: we become aware that we are renegading upon the very covenant with our common Creator.

In this situation we are driven to make amends, to attempt to redress the balance, back onto 'the side of the creation'. We become aware that mankind's current powers, borne by industrialisation, not only give mankind a new control over nature, but also impose upon mankind new responsibilities towards nature. It is this drive, in the first place, which is at the root of the modelling of the natural environment.

The problems of the environment have led to the social phenomenon of the environmental movement. The aims of this movement are the traditional ones of any social movement: of making us aware of its problems, of appealing to our inborn senses of responsibility and justice to take up these problems and to make them our own, and of translating the social pressure that results from this taking-up process into political action. This movement draws upon our most basic aesthetic and ethical intuitions as a means of redressing the balance of nature. These aims of the environmental movement cannot be realised, however, without the means to bring us to a proper state of awareness and, further, to a properly-founded sense of personal and collective responsibility. The aim of numerical modelling, in the first place, and of hydroinformatics, more generally, is then to assist in inculcating the necessary state of awareness and of making clear and definite our own personal and social responsibilities. The aim of hydroinformatics is to confront us with the consequences of our actions, and in this respect with our actions towards the world of nature. First numerical modelling, and subsequently hydroinformatics, then become means of 'challenging-out' the individual and society to confront his, her or its own Either-Or: the hydroinformatics system tells us that if we do *this*, a certain consequence must be expected, and if we do *that*, another consequence, and so on again. After that, we must choose between these consequences: Either-Or. Thus the aim, first of computational hydraulics and then of hydroinformatics, is to make the ways of our world so transparent that we may proceed into the future of this world with *concernful circumspection*, and thereby we may proceed in what we judge, to the very best of our abilities, to be the right direction.

First the numerical model, and subsequently the hydroinformatics system generally, are thus used in this area in order to obtain a certain *ethical effect*. This effect is best achieved, however, through the use of the most *aesthetic means* possible. The aesthetic itself has two sides. The first of these is the aesthetic of the system builder, which we may call the *inner aesthetic*. The second is that which appeals to the user of the system or the system results, and this we shall call the *outer aesthetic*. These aesthetics are related through the system itself, which in the case of the model or domain-knowledge encapsulator is a set of numbers imbued with a certain logic.

2.1.2   Aesthetic deception as the primal form of outer aesthetic

In Abbott and Basco (1989), numerical models are introduced as number allegories. However, when seen from the point of view of most individuals in our current societies, who are not initiated into the symbols, conventions and rites of numerical modelling, these models must appear even more as *fantastic* number allegories. After all, we are asking these individuals to believe that one particular collection of numbers is a realistic description of the circulations of the waters of the northern Polar Basin, and that another — and for most individuals indistinguishable — collection of numbers is a realistic description of the spreading of an antibiotic in the human eye. Indeed, we claim that the art of numerical modelling resides precisely in our ability to 'read meanings' into sets of numbers, which claim must stretch the credulity of most individuals to the limit.

It is for this reason that, already today, the end-users of the model — those whose ethical drives have to be awakened and sharpened — are hardly ever confronted with the numbers themselves, but nearly always with some graphical representation of the set of numbers. We are nowadays all familiar with colour plots and CRT screen graphics of the type already introduced in this book, as visual representations of 'the information contained in the set of numbers'. Of course, the shading and the colour, the projection in two-dimensions and the intimation of three-dimensionality by projection in perspective, the smoothness and, increasingly, the motion, are all *deceptions* of the visual senses, whereby the information contained in a set of numbers is transformed into a form that we can more readily assimilate. We have in fact to do with a *deception of the senses*, or an *aesthetic deception* (the Kierkegaardian *Sandsebedrag*), as the means of imposing an ethical condition (Kierkegaard, 1859). This is indeed the general way of ethical action of the outer aesthetic. However, as with all such means of deception, this particular approach can be used both to enlighten and to mislead. The 'Great Graphical Illusion', as and by itself, is a very dangerous thing (Abbott, 1986). It may provide very realistic-looking results — but are these 'really realistic'? The answer to that question brings us back to the modeller, and then further to the aesthetic of the modeller, which we have already introduced as the inner aesthetic.

2.1.3   The logic behind the numbers as the primal form of inner aesthetic

Chorin (1983), even while speaking from the standpoint of a more scientific computational fluid dynamics (CDF), expressed one central aspect of the inner aesthetic, as this functions already in second- and third-generation modelling, as follows:

> Just as there is a notion of mathematical elegance that is difficult to explain to non-mathematicians, so there exists a separate notion of 'beautiful calculation' that is hard to explain to the uninitiated: roughly speaking, a beautiful calculation is one in which mathematical constructions interact in interesting ways with reality, or one in which a complex physical phenomenon is explained in an unexpected and inexpensive way.

> Correspondingly: CFD, like quantum theory, requires new, broader and more complex notions of what it means to understand phenomena.

It follows again from the very presence of an aesthetic in modelling that modelling itself cannot be entirely scientific. Another kind of intuitive, more artistic understanding has then to be mobilised alongside scientific understanding, and since the time of Dilthey (*e.g.* 1976) it has been usual to extend the range of application of *hermeneutic understanding* to cover this broader kind of understanding. The successful modeller is then one endowed with hermeneutic understanding, over and above scientific understanding. Similarly, in the words of Leendertse (1981):

> The way a modeller derives a model for the system he is studying can best be described as an intuitive art. No fixed rule is given. The modeller must have the ability to analyse a problem, abstract its *essential* features, select and modify assumptions that characterise the system, and subsequently

extend and enrich it until a useful approximation is found...
Moreover ... the modeller is at least as important as the model
which is used.

Thus, the inner aesthetic of modelling is an aesthetic of the model and
the modeller indivisibly. This aesthetic has then to do both with the standard-
scientific truth relation of the model to *its* prototype in the natural environment
— of which the modeller may have only a partial and incomplete view at any
one time — and with the truth relation of the modeller to his or her outer world
of common-sense experience of the prototype in the natural environment. This
relation is schematised in Figure 17.

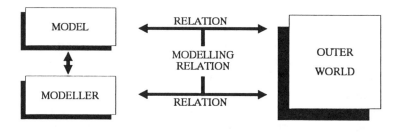

Figure 17   **Schematisation of the 'relation between relations' of numerical
modelling**

The relation of the modeller to his or her outer-world prototype has
itself a relation to the relation of the model to its prototype. However, every
relation can be expressed in terms of a logic. Behind the relation between the
modeller and the outer-world prototype of the modeller there is a logical
structure, but alongside this we postulate the existence of a logical structure that
underlies the relation between the model and its prototype. When seen from this
point of view, the activity of modelling is essentially one of a working-out of
these two underlying logical structures and of making a unity of them. This
working-out itself proceeds through a process of interactive working between
the modeller and the model, which process is again not entirely scientific, so
that it is not itself entirely logical in the standard-scientific sense. The model can
be said to 'learn' from the outer-world experiences of the modeller, but, just as
surely, the modeller 'learns' from the behaviour of his or her model. What the

modeller and the model ultimately together 'learn about' is, of course, the natural environment.

In order to exemplify this notion of a 'relation between relations' we may consider the phenomenon of *instability*, as this occurs in numerical models. Instability is said to occur in a numerical model when the magnitudes of the numbers generated by the model increase without bound, so that these magnitudes come to exceed the capacities of the registers of the digital machine, thus causing this machine to abort the computation. This is discussed more in an appendix to Chapter 3. It will suffice here to observe that it has been argued (Abbott and Basco, 1989) that *instability is the numbers' way of telling us that our model contains contradictory statements*.

It then follows at once, however, that numerical instability must be of the greatest possible value in the development of a naturally-realistic numerical model, for its intercession immediately indicates that our model contains internal contradictions and we can run a series of tests in order to localise and so identify the contradiction concerned. This in turn indicates where internal contradictions arise in our own minds, as the minds of the modellers, in our description of our natural environment. It is this insight into the essential nature of instability that lies behind the maxim that one should always develop a model at the outermost limits of stability, for it is in this way that the internal contradictions in our own thinking about our environment are most clearly and immediately exposed by our model. Accordingly, it is this approach that is followed almost exclusively in the development of models and modelling systems.

Now the most remarkable property of numerical instability is that we are *forced* to listen to what the numbers have to say. It will avail us nothing to refine our model description so long as we persist in our inner contradiction: the instability will only make itself manifest earlier and even more forcibly. *Instability is then a kind of 'sickness unto death' of numerical modelling.* Reflecting its Biblical prototype (John XI, 4), and its Kierkegaardian representation, numerical instability can be seen as the property of sets of numbers and sets of operations on sets of numbers (which sets of operations again have number representations) which expresses the model's, and thus our own, state of internal contradiction. In the modelling process, the model serves as a sort of 'truth-finding device' for the state of our own mind *vis-à-vis* our natural environment, whereby whenever we introduce an own untruth into the model, the model 'blows up', and necessarily to our own total frustration. The 'despair' of the model, whether 'because it cannot be its true self' (Kierkegaard's α or 'feminine' form of despair), or just 'because it is its (still untrue) self' (the corresponding ß or 'masculine' form), becomes our own personal despair — an experience that will be familiar enough to practical modellers! In this way the model might be said to provide an 'ethical control' over the modeller's otherwise aesthetic extravagances, and it is in this way that modelling teaches us something about our own selves. This is another and ultimately deeper way of seeing how 'the numbers have begun to function in another way'.

Thus, to summarise, the mobilisation of the ethical proceeds through the inner and outer aesthetics of the modeller, where the corresponding 'model' should now be taken in its wider, hydroinformatics sense. The power of the aesthetic of the work of art is that it propels us to an intimation of the essence of 'the thing in itself' (the Kantian *Ding an Sich*). The power of the outer aesthetic of hydroinformatics, on the other hand, is that it propels us to a sense of our own relation and our own responsibility to the 'the thing in itself'. Alongside this aesthetic (and inseparable from it), hydroinformatics also has an aesthetic in the first sense, of the experience of the work of art — which is what we have called its inner aesthetic; it is only that the intimation is in this case of a quite other kind than that occasioned by the experience of the finished work of art. In hydroinformatics, the experience is nothing like so universal or immediate; indeed, it can only be experienced at all by those initiated into the development of the innermost workings of the hydroinformatics system. It was in this way that Leendertse's characterisation was particularly apt, when he describe modelling generally as an *intuitive art*.

Before concluding this general discussion of aesthetic drives, however, let us return to our first characterisation of hydroinformatics and of 'the fact that now, whenever we try to point to modern technology, the words setting-upon, ordering, and standing-reserve, obtrude and accumulate in a dry, monotonous, and therefore oppressive way.' We now see how this first impression of hydroinformatics, as such a 'dismal science' of the ordered, the countable and the computable, tends to be *inverted* by the deeper, or more abstract, view, so that hydroinformatics comes to appear in a much brighter light. We see here already how hydroinformatics becomes itself a field in which opposing forces come to contend: on the one side the subjection of the whole world of nature to the status of a standing reserve, and on the other side the means to mobilise a new solicitude towards this same complete world of nature.

To these general remarks, some more specific reference should be made to the nature of this logic that lies behind the numbers. After all, we have just been emphasising the non-scientific aspects of modelling, and the introduction of a characterisation of numerical instability that is more 'dogmatic' (*i.e.* based on repetitive observations) than scientific. We can perhaps best introduce this other kind of logic from the side of AI, where Kay (1984) spoke of the non-scientific aspects of modelling as follows:

> What does it mean to represent something from our world in a
> world that is not of our world? What does it mean to do things
> in that world that are not like the things in this world? How can

we translate back into our world in such a fashion as to get a message that actually means something to us? I think that is mysterious to everyone. Most people in Artificial Intelligence who have had a glass of wine will tell you they do not actually understand it much either. The reason for the mystery is that the correspondences between what we think is going on in the world and the symbols that we use in communication are very fragile. The nature of much of the fragility is not understood. Every time we make a model of some kind, we leave out lots of things, and we base what we do on guesses about causal relations rather than on firm understanding.

Even more generally and critically, Shackle (1967, pp. 157-160) has described the impact of Keynes' *General Theory* upon the illusions of pre-computational, 'applied-mathematical' modelling in economic theory, as follows:

For in these models each variable has, in effect, its own determining equation, each in turn is exhibited as dependent on some of the others, and we have an insulated, closed and complete set of a very few variables mutually determining (once some 'initial' values are given) the skein of time-paths they shall follow. For Keynes, by contrast, there are economic wind and weather, in the form of politics, invention, fashion and the incalculable movements of expectation, great forces outside of and unshaped by the economic ship whose course we hope to understand and control. ... Just in what sense variables are variable, when a governing environment is assumed to be given and specified; who or what varies them; these and such matters are deliberately left as visible but unanswered questions. The reader himself must think and must elect his solution or hold several interpretations in mind. This situation is made clear to the reader and imposed upon him, with superlative skill and tact. The ease and grace of the style itself prepares the reader for the need to seek more harshly exact instructions. ... The values of these variables are, must we assume, the very jetsam of the tides of history in all its depth and complexity; to seek to 'explain' them would be to trespass far beyond the bounds of technical concern which the economist sets himself, and compel him to claim competence where he, and any man, can have none. History herself varies these variables, by an arcane process whose nature we shall do well, for our own purposes, to leave inviolate.

... In the end, Keynes speaks of the 'ultimate atomic independent elements'. What can 'ultimately independent' mean, except either that they are too complex and elusive a fashion for us to penetrate, or else indeed they are the upshot of something spontaneous and originative, or 'random', at the very source of history? Keynes does not tell us.

Thus, behind all the things which we do understand in a modern-scientific way (which are the nominal subjects of modern science), we find a number of other things which we cannot understand in a modern-scientific way (and which are correspondingly 'non-scientific', at least in a modernist ideology). In particular, to the extent that we admit the intervention of non-scientific aspects into our modelling with sets of numbers, so we have to admit that our 'logic behind the numbers' has, also, non-scientific attributes. But what kind of logic is this, which has non-scientific attributes? The simplest answer would be that it is 'the logic of our own minds', and indeed just this characterisation has been used to justify the elaboration of so-called 'neural network computers', it being claimed that these are also 'not entirely logical' devices that simulate certain life-essential processes that are generic to the human mind. We shall return to this matter in Chapter 3.

2.2   The ecological dimension of hydroinformatics

2.2.1   The problem of values in ecology

Hydroinformatics is the name of a new possibility that opens up other new possibilities again. One of these other new possibilities is that of the control of our aquatic natural environment, in the manner discussed in the previous chapter. However, as already introduced, this control cannot now be regarded only as a means of *reducing* the world of the waters to a standing reserve for our partial, locally-perceived interests; it now becomes much more one of *mobilising a new solicitude towards this world of the waters*. This task of mobilising a new solicitude towards the world of the waters then constitutes a movement into the *ecological dimension of hydroinformatics*.

Let us first observe that this dimension must at first appear as a very different one from that of the control of our own natural environment. Just to begin with, we now have to face the question of what is our 'own'. From the common-sense point of view of fishermen, it is obvious that all the fish in the sea are our own, just as much as it is obvious to the common sense of a shepherd that all the sheep in his flock are his own. From this point of view, we place no value at all upon the needs of our natural competitors in the respective food webs, such as, in this case, seals and wolves. Indeed this position is currently becoming entrenched in the overall objective of 'sustainability', where interest is focused on the ability of ecosystems to sustain a maximum throughput of heat energy and chemical and biological 'wastes' while still satisfying certain common-sense social-economic requirements. The inference must then be the obvious one of common-sense reasoning, that the ecosystem should be sustainable *for us*. From the common-sense point of view, this is the only *value* that can be placed upon an ecosystem altogether: that it is *for us* in the most narrow and immediate sense. However, if the word 'ecology' is itself to retain its original meaning at all, such common-sense interpretations cannot be sustained. Indeed, we cannot proceed at all along the ecological dimensions of ecology if we try to maintain values of this kind.

Thus the *first problem* of hydroinformatics when it seeks to acquire an authentic ecological dimension is one of acquiring corresponding *ecological values*. However, this very problem, which is essentially one of the recognition and thus the attainment of the 'highest values', has long appeared as *a central problem of modern philosophy*, and at the very least since Nietzsche (Heidegger, 1961, p. 61):[4]

Alle Werte, mit denen wir bis jetzt die Welt zuerst uns schätzbar zu machen gesucht haben und endlich ebendamit *entwertet* haben, als sie sich als unanlegbar erwiesen — alle diese Werte sind, psychologisch nachgerechnet, Resultate bestimmter Perspektiven der Nützlichkeit zur Aufrechterhaltung und Steigerung menschlicher Herrschafts-Gebilde: und nur fälschlich *projiziert* in das Wesen der Dinge. Es ist immer noch die *hyperbolische Naivität* des Menschen, sich selbst als Sinn und Wertmaß der Dinge *anzusetzen* ...'

That this characterisation is correct of 'common-sense truths' would be little contested today, and indeed it has become an everyday one within the

---

[4]I have been unable to obtain translations of this and some other works used in reference. In each such case I have left the original in the main body of the text and provided my own translation. In this case: 'All the values that we have so far introduced in order to make the world so estimable, and then finished up by devaluing when they showed themselves to be untenable — all these values are, when worked out psychologically, really only results of our taking definite perspectives on the usefulness of raising up and magnifying an image of our own domination; and these values are only falsely *projected* as being of the essential nature of things. It still belongs to the *extravagant naivity* of man that he puts himself up as the purpose and the measure of everything.'

'Green' movement. However, it applies not only to common-sense truths, but also to modern scientific truths as well. This was expressed with exceptional perspicuity by Husserl (1938/1973, p. 42-44), as follows:

> The world in which we live and in which we carry out activities of cognition and judgement, out of which everything which becomes the substance of a possible judgement affects us, is always already pregiven to us as impregnated by the precipitate (*Niederschlag*) of logical operations. The world is never given to us as other than the world in which we or others, whose store of experience we take over by communication, education, and tradition, have already been logically active, in judgement and cognition. And this refers, not only to the typically determined sense according to which every object stands before us as a familiar object within a horizon of typical familiarity, but also to the horizon-prescription (*Horizontverzeichnung*), the sense with which it is pregiven to us as the object of possible cognition, as an object determinable in general. The sense of this pregivenness is such that everything which contemporary natural science has furnished as determinations of what exists also belongs to us, to the world, as this world is pregiven to the adults of our time. And even if we are not personally interested in natural science, and even if we know nothing of its results, still, what exists is pregiven to us in advance as determined in such a way that we at least grasp it as being in principle scientifically determinable. In other words, for this world which is pregiven to us, we accept the following idea as a matter of course on the basis of modern tradition, namely, 'that the infinite totality of what is in general is intrinsically a rational all-encompassing unity that can be mastered, without anything left-over, by a corresponding universal science'. This idea of the world as a universe of being, capable of being controlled by the exact method of physicomathematical science, of a universe determined in itself (*an sich bestimmten*), whose factual determinations are to be ascertained by science, is for us so much a matter of course that we understand every individual datum of our experience in its light. Even where we do not recognize the universal binding force and general applicability of the 'exact' methods of natural science and its cognitive ideals, still the style of this mode of cognition has become so exemplary that from the beginning the conviction persists that objects of our experience are determined in themselves and that

the activity of cognition is precisely to discover by approximation these determinations subsisting in themselves, to establish them 'objectively' as they are in themselves — and here 'objectively' means 'once and for all' and 'for everyone'. This *idea of the determinability 'in itself' of what exists* and hence the idea that the world of our experience is a universe of things existing in themselves and as such determined in themselves is so much a matter of course for us that, even when laymen reflect on the achievement of knowledge, this 'objectivity' is from the first accepted as self-evident.

In relation to ecology, the most immediate consequence of this attitude is seen in the use of the word '*ecosystem*'. For, as has now been made abundantly clear by many authors, the whole notion of 'system' is a local and contemporary one, peculiar to the 'self-centred' and so anthropocentric mode of modern thinking. For example (Heidegger, 1977, p. 141):

In the Middle Ages a system is impossible, for there a ranked order of correspondences is alone essential, and indeed as an ordering of whatever is in the sense of what has been created by God and is watched over by His creature. The system is still more foreign to the Greeks, even if in modern times we speak, though quite wrongly, of the Platonic and Aristotelian 'Systems'.

However, beyond this again, 'system' is synonymous with 'mathematisable' — the potential to be expressed in mathematical form — and indeed it points to a very definite approach to mathematics itself, and one which was criticised in considerable detail by Wittgenstein (*e.g.* 1969, 1975; see also Diamond, 1976). Thus, even as we proceed along the ecological dimension of hydroinformatics, we come to experience 'ecological values' which put into question all the tenets introduced in the first two cycles of our characterisation of hydroinformatics. Moreover, as Wittgenstein made quite clear, this difficulty cannot be even so much as *posed* as a mathematical problem. This, then, is the first problem of hydroinformatics when it seeks to acquire a truly ecological dimension.

There have of course been many attempts to solve this problem, or at least to alleviate its worst consequences. Nietzsche proposed a multi-faceted approach to the viewing of the phenomenon itself (Heidegger, 1961, p. 167):

'Objektivität', — letztere nichts als 'interesselose Anschauung' verstanden (als welche ein Unbegriff und Widersinn ist),

sondern als das Vermögen, sein Für und Wider *in der Gewalt
zu haben* und aus- und einzuhängen: so daß man sich gerade die
Verschiedenheit der Perspektiven unter der Affekt-Inter-
pretationen für die Erkenntnis nutzbar zu machen weiß.

... Es gibt *nur* ein perspektivisches Sehen, *nur* ein perspektivisches 'Erkennen'; und *je mehr* Affekte wir über eine Sache zu
Worte kommen lassen, *je mehr* Augen, verschiedne Augen wir
uns für dieselbe Sache einzusetzen wissen, um so vollständiger
wird unser 'Begriff' dieser Sache, unsre 'Objektivität' sein.[5]

Thus, on this basis, there can be no one, neutral model of an eco-
'system', but there can only be a number of different models, corresponding to
a number of different eco-'systems', each of which corresponds to another
perspective, another emotional interpretation. We must then multiply the number
of our perspectives — which then however suggests, in its turn, that we will
only succeed in arriving in another 'box' again: within an eco-'metasystem'. As
we shall subsequently argue, the essential point here is that we extend, and
indeed totally transform, what we understand by a standing reserve, and so what
we understand by 'common-sense truths'.

We need not go into other attempts at a solution (some of them more or
less scientific, others downright anti-scientific) which have been proposed over
the last century or so, and many of which are still in circulation (see, for
example, Bohm, 1980, Hoyle, 1980, and Capra, 1978, for those with at least
a scientific background). It will be more productive to proceed at once to the
second problem that arises when we attempt to move into the ecological
dimension.

In apparent contrast to the first problem, the second appears to be a
problem of a *technological* nature. Being now forewarned by Heidegger's
analysis of the essence of technology, however, we will already be very cautious
about relegating this problem to any kind of 'second place' on the basis of such
an outer appearance. The second problem has to do with the fact that ecology

---

[5]*Author's translation:* Objectivity, this last not to be understood as 'disinterested observation'
(which is, as such, inconceivable, a pure nonsense), but as the ability to control the giving of
evidence, both for and against and both from inside and from outside; so that there is a precise
understanding of how to make *differences* in perspectives, occasioned by differences in emotional
interpretations, useful to the understanding.

Things can only be seen at all from a particular perspective, only understood at all from
a particular point of view; and *the more* the kinds of emotions that we allow to come to
expression about a particular case, *the more* pairs of eyes, different pairs of eyes, through which
we are able to see this case, the more complete will be our 'conception' of the case, and the
greater will be our 'objectivity'.

studies, or at least should be studying, the interactive coexistence of *intelligent beings*, which beings accordingly have their own needs and rights alongside our own perceived needs and rights to these beings as our own 'standing reserves'. The second problem is then that of describing this interactive coexistence of intelligent beings, and then in hydroinformatics, with the aim or object of protecting, and possibly further extending, our own standing reserves. Indeed, to the extent that we attempt to come to solve our first 'problem', we *must revalue our whole concept of 'standing reserve' so as to include our solicitude towards nature within it.*

The second problem of hydroinformatics is then one of describing an interactive complex of intelligent beings, subject to fluid advection, diffusion and other flow processes, and with a view to serving certain interests that we now identify much more generally and directly as 'our own'. However, the part of the problem that has to do with modelling intelligence is the specific field of study of Artificial Intelligence. Thus our movement into the dimension of ecology must bring us back again into AI.

We are now brought back to AI, however, *in a new way*, in that we are now dealing with organisms that have other modes of intelligence than our own, and other 'aims' and 'objectives'. Of course, it is still we who are reading these 'modes of intelligence', 'aims' and 'objectives' into our observations of the behaviour of the organisms, so that our description must be in a form that we can ourselves most easily relate to our observations. But still we must, of necessity, regard AI differently than in its 'standard', anthropocentric way. In particular, we must understand how we are imposing a language-structured intelligence upon the behaviour of beings whose intelligence is structured in quite other ways.[6]

---

[6]There is something very strange here, in that in the four standard aquatic- and marine-ecology texts used as background reading for this study (Barnes and Hughes, 1982; Mann, 1982; Wetzel, 1983; Moss, 1988), the notion of 'intelligence' scarcely figures at all. From the present point of view, however, *an ecosystem is a consequence or a manifestation of interacting intelligences*, so that every such systematisation, and so every ecological model, must make use of AI. Although some studies have an inclination in this direction (see, for example, Mann, 1982, pp. 273-277), this does not seem to be at all widely understood. (Correspondingly, and to close the hermeneutic circle, intelligence is itself defined as any strategy of action and inaction that promotes the survival of the phenotype, so that intelligence has itself an ecological ground).

In the same vein, among the six of the fifteen volumes of the Elsevier series on *Developments in Ecological Modelling* used as background reading in preparing this piece, only that of Jørgensen (1986) extends the standard Lotka- and Volterra-like modelling approaches by introducing logical relations between components through graph-theoretic devices (and again using an adjacency matrix, in this case taken from Halfon, 1983). But even here, although the possible introduction of KBS technology is mentioned, this is treated separately, and the connection between these methodologies is not made.

More generally, we see again in these standard works the truth of Heidegger's insight, introduced in the very first section of this book, that modern science tends to view the whole

Now, to some extent, this new way has already made its appearance within AI. For example, in his paper 'Ascribing Mental Qualities to Machines', McCarthy (1979) has shown (in the words of the summary of Genesereth and Nilsson, 1987, p. 8) 'how *convenient* it is to talk about artefacts (such as thermostats and computers) as having mental qualities (such as beliefs and desires). For example, according to McCarthy, a thermostat *believes* it is too hot, too cold, or just right and *desires* that it be just right.'

However, as we reassess AI from this point of view, we see how the entire subject has developed as a model in which the point of view expressed by McCarthy has always been latent. We can perhaps best see this by starting 'at the top end' of AI, with the notion of an intelligent agent and associated architectures of interacting intelligent agents. Thus, keeping to Genesereth and Nilsson (1987, p. 211):

> This semantic groundwork is based on the supposition that each reasoning agent in the world has a theory (about the world, say) composed of ordinary, closed wffs ['well-formed formulas'] and closed under that agent's deductive apparatus, Note that we do not assume that an agent's theory is closed under logical implication: it is closed only under the inference rules of that agent. An agent may have an incomplete set of inference rules, and therefore it might not be logically closed. The distinction is important if we want to reason about agents that themselves may have limited reasoning power.

We then find that the very vocabulary is evocative of the world of nature, such as (Ibid, p. 309):

> A *tropism* is the tendency of an animal or plant to act in response to an external stimulus. In this section, we examine a class of agents, called *tropistic agents*, whose activity at any moment is determined by their environment at that moment.

---

world merely as a standing reserve, as *Bestand*, as something existing only in and for our own use. From such a point of view, the organisms only enter in the forms of biomasses, energies or other extensive variables, and nowhere is it even acknowledged that these organisms possess, in a fully integrated manner, senses, inferential capabilities, strategies, modes of locomotion, clocks for feeding, reproducing and other activities, and much more besides; and then they possess, beyond all this again, what we can only describe within our current ideology as 'life codes' and 'death codes'. The Lotka-Volterra, or *Bestand*, school of ecology must then naturally alienate many biologists and others, who see in their fellow organisms the highly-refined products of millions of years of evolution, or, which is exactly the same thing, the products of the creation itself.

Then, later referring back to this (p. 311):

> The agent introduced in the last section is extremely simple.
> Since it has no internal state, it is forced to select its actions
> solely on the basis of its observations — it cannot retain
> information about one external state for use in selecting an
> action to perform in another state. While there is no need for an
> internal state in this simple case, the ability to retain information
> internally is extremely useful in general. In this section, we
> specialise our definitions, from the last section, to cover agents
> with internal state, hereafter called *hysteretic agents*.

Thus AI has already occupied itself over some decades in constructing
a conceptual and machine-logical apparatus to describe, at least partially and
incompletely, interacting, intelligent agents, and it has thereby modelled, albeit
partially and incompletely, the world of nature. There is already a definite
relation between our view of an intelligent being and our construction of an
intelligent agent. The objection can at once be raised that this AI 'machine' is
based upon our own transpositions of our own logical processes into ecological
processes, but even here AI has developed its 'neural-networking' approaches,
which appear to reduce the extent to which we dictate the logical connections
directly, and which we shall discuss in the next chapter (*e.g.* Rumelhart and
McClelland, 1988). Furthermore, AI has learnt to deal with uncertain inference
at its own scientific level, whether by probabilistic means or by the introduction
of 'fuzzy' sets. (See, for example, Genesereth and Nilsson, 1988, Chapter 8.
By way of historical background on the fuzzy-set side, see Carnap, 1937, *e.g.*
pp. 239-247. On the side of probability, reference must similarly be made to de
Finetti, 1974: see also, in this respect, Jeffrey, 1984).

Even so, it has to be admitted that from the point of view of
hydroinformatics the existing apparatus of AI is highly unsatisfactory, in that it
is not concerned with the locations and densities and physical activities of its
intelligent agents in time and space: it does not deal with their physical
interactions, such as their eating and their being eaten — with their
corresponding changes in biomass, their hunger and their reproductive needs,
their social structures, including age distributions, and so on. It does not take
account at all of what Husserl called the 'life worlds' of the individual
organisms. Of course, so far, putting aside neural networks, all of this can
appear in hydroinformatics only in the form of sets of *quantities*, represented by
sets of *numbers*, as these are influenced by fluid processes (advection, turbulent
diffusion and turbulent dispersion, shear flow structures, upwellings, etc.) again
described by sets of numbers. This does none the less mean that, in
hydroinformatics, the intelligent agents have now themselves to be given a

spatial and temporal 'life world' describing matrices of food webs, breeding patterns, age distributions and any number of other physically-observable and quantifiable features of 'the' ecosystem, and this has to be further embedded in a matrix of fluid flows, with all their transport capabilities (*e.g.* Abbott and Warren, 1974; Abbott *et al*, 1977).

Thus we again see why, in hydroinformatics, thc logical apparatus of AI modelling has to be integrated with the computational elements of modelling, while, quite symmetrically, the 'ways of the numbers' have now to be influenced in new ways by the 'ways of the intelligences that steer the numbers'. This is necessary — even as, as will be explained in Chapter 3, it is still almost certainly insufficient.

We can thus, even on this limited basis, return to the first problem of hydroinformatics as it starts to explore its ecological dimension. For if the numbers are beginning to function in another way, is it not possible that the so-called 'logics' to which they are so closely connected may also come to function in another way than heretofor? Are we not ourselves going to learn some lessons from what are, nominally at least, our own artefacts? Are we not going to arrive at another view of what we meant by this: *for us*? Indeed, to complete the first cycle of this chapter, are we not going to arrive at a new solicitude towards the world, and thus a new 'life world' of our own? We can better leave this discussion to the end of this work, however.

There is a third problem, which in its turn opens up a next cycle, but it must suffice here only to pose the problem. It is that hydroinformatics cannot explore its ecological dimension without the participation of a considerable number of individuals of very varying backgrounds. The corresponding 'hydroinformatics system' must then have a structure (such as, in the function-orientated paradigm, a 'frame') that allows for a great variety of contributions, provided by the great variety of backgrounds and concerns of contributors (such as, in the case of a frame, by providing 'slots' for these contributors). At the very least, this makes necessary the use of a 'list-processing language' (*e.g.* LISP) that can comprehend lists of lists, etc., and in which items on the lists can finally be either programs or data. However, in the real world of economic contraints, hydroinformatics will have to proceed way beyond the writing of programes directly in LISP, so as to provide a suitable variety of environments to the variety of kinds of individuals who are contributing to its 'systems'. This is not so much a problem of computer science *per se*, but it is more a problem of making use of certain products of computer science (including AI), often in ways that were not envisaged by the originators of these products. The challenges presented to hydroinformatics can only be met by resorting to new uses of general products in computer sciences, and of AI in particular. Very generally, the need here is to introduce systems that are 'open' to their users: these will be much less specific 'tools' and much more 'environments' in which chemists, biologists, zoologists, sedimentologists and others can themselves build and modify tools. Obviously, then, these 'metasystems' must include facilities for truth maintenance and more general means for accommodating or avoiding conflicting views.

It should further be mentioned that a movement in the ecological dimension of hydroinformatics must have considerable repercussions on hydraulics, and, in particular, on computational hydraulics. This is because many interactions between organisms are sensitive to a variety of fine-scale flows, such as are currently grouped together under the one rubric of 'turbulence'. The study of ecosystems will necessitate a more detailed description of fine flow structures, of processes of turbidity and of mechanical and thermodynamical interactions between solids, gases and waters generally. Thus, the challenge of marine and aquatic-ecological modelling has to be passed on, at least to some extent, to computational-hydraulic modelling.

If, in conclusion to this section, we consider the current situation, marine and aquatic, in ecological modelling, we observe that, for the most part, this has not come much further than function-orientated, frame-and-slot, modular programming of the type that was already applied consistently in third-generation modelling in hydraulics and hydrology from the 1970s onwards (such as in the SHE: see again Abbott *et al*, 1985). Thus, under the European Commission's MAST programme, to be implemented in the 1990s, the proposed European Regional Seas Ecosystem Model (ERSEM) is a modular generic ecosystem model. No mention whatsoever is made of concepts of AI in the MAST documentation, even though at least one of the participants in ERSEM has mentioned this possibility elsewhere. Neither does there appear to be any AI modelling component in the Commission's Science and Technology for Environmental Protection (STEP) programme. On the other hand, the application of an an early, relatively low-level software environment (IBM's *Continuous System Modelling Program*) was explored by Radford (1979), building further on earlier second-generation modelling (Radford, 1971). More recently, Baretta and Ruardij (1988) have described a second-generation environment customised to the Ems and western Wadden-Sea areas.

On the side of the specific ecosystems, a considerable number of generic models have, of course, been constructed, but then few of these attempt to take an overall ecological view (*e.g.* Kremer and Nixon, 1978; Radford, 1979; Baretta and Ruardij, 1988; de Vries *et al*, 1988). In by far the greater number of studies conducted so far, only a few system processes have been researched in local detail, while many other essential system processes have hardly been considered at all. For example, although much has been done on horizontal variations in ecosystems, only a few models describe the vertical structures of ecosystems (Kiefer and Kiefer, 1987; Stigebrand and Wulff, 1987; Radach and Moll, 1989) even though the most significant differentiations of ecology itself (pelagic, benthic etc.) are defined in terms of variations in the vertical. In the case of energy and related biological information flows through the ecosystem, this appears exceedingly unsatisfactory. (See, for further emphasis of this point, Joint and Morris, 1982, Azam *et al*, 1983, and Smetacek and Pollehne, 1986, on the nature of 'microbial loops' in ecosystems.) This indifference to the computational-scientific and other exigencies of complex-system simulation modelling is observed also in data modelling. Apart from a few adaptions of relatively minor standard proprietorial SQL data bases and some adaptations of Geographic Information Systems (GISs), very little appears to have been done.

On the whole then, the ecologists are, by the very nature of their problems, driven towards a more 'holistic', truly-ecological view, so that they increasingly aspire to such a view, but very few of them have yet seen the information-technologic implications of their aspirations. For the most part, they have not yet become aware of the explosion of 'if-then' statements and associated loops, triggers, counters and other control elements that must be invoked unless they make a quantum leap in control structures. The very magnitude of their ambitions (depth differentiations, interacting species and genera, etc.) makes for an immense control problem, over and above the data-base, knowledge-base and operational difficulties entailed in realising such ambitions.

One of the most immediate objectives of hydroinformatics is to instill an awareness of this problem, and of its possible solutions.

2.3   The third-world dimension of hydroinformatics

The 'life world' that imposes its outer form upon hydroinformatics is that of 'ourselves'; but who or what are the *us* and the *we* to which reference is then made? Surely the most superficial glance at the human populations of our planet Earth must show that there is no one *us*, no one *we*, but there are very many of these '*us's*' and '*we's*'. Indeed, at the largest scale we commonly talk about first, second and third *worlds*, even though as at this time of writing the second of these appears to be breaking up into two parts, the one of which appears to be gravitating towards the first world, the other perhaps towards the third. Of course the ecological dimension remains active in all three worlds, but it must be extended into another dimension again, which is that of civil society, and the way that this extension occurs differs from the one world to the other.

Just as the dimension of ecology may be described as the dimension of wholeness, of oneness, so this extra dimension, into which the ecological must itself be extended, is characterised by its very nomenclature: it is itself of the orderable, the countable and the computable, but now operating at the largest scales of human society. This is, then the dimension of what we have here come to call the 'dismal science' in society itself: ultimately, it is the dimension of *despair*. Since Kierkegaard (1849) it has been clear that the dimension of despair takes no cognisance of material wealth, although material wealth does provide more ways of drowning-out the voices of this despair. (For example, at this time of writing, the second largest industry of all, after armaments, is that of chemical narcotics, with 'entertainment'— *i.e.* primarily audio-visual narcotics — not very far behind). Of course, hydroinformatics, *per se*, can have nothing at all to say about despair itself: this 'sickness unto death' of the spirit (which we have already introduced and to which we shall subsequently return) can only influence hydroinformatics directly to the extent that it is able to reduce hydroinformatics also to a 'dismal science'. The overt despair of the third world makes us aware of our own, otherwise covert, despair, and it is this which causes us to act to alleviate what we perceive, in its outer manifestation, as extreme poverty and accompanying misery. This state of despair has then everything to do with the social applications of hydroinformatics, and especially and most clearly with its applications in the third world. After all, the *potentialities* of hydroinformatics are unlimited in terms of improving the material conditions of life of third-world societies, and in this respect hydroinformatics is by no means unique; but, *so far, in practice*, such 'high-technology' applications have proved exceedingly difficult to realise, and, even when realised, they have usually achieved only very partial successes — and then not for any particular 'technical' reasons, but precisely because of the sort of grip that despair itself exerts within third-world societies.

It follows from this that in its implementation, and especially in the third world, hydroinformatics has to come to terms in one or the other way with the dimension of despair. In fact such a necessity is already conceded in the elaboration of many water-resources projects, where it is accepted that these projects frequently give rise to marked changes in social structures, many of them exceedingly detrimental to overall social welfare. This is now reflected in the terms of references of water-resources projects, such as those of the World Bank, which include a consideration of the social consequences of projects (see, for example, Hoogstraten, 1985; White and Gordon, 1987; IRC, 1988; and Schramm and Warford, 1989). Indeed this aspect already enters into works on ecology itself (*e.g.* Moss, 1988, pp. 2-9, 125-130, 333), where it is accepted that the 'ecosystem' must itself take cognisance of human social interaction. This is to say, however, that the hydroinformatics system must now incorporate facts and rules of a sociological nature, and that these systems must be accessible, or 'open' to social anthropologists, other sociologists and others again who are concerned with these matters.[7]

This in turn implies that hydroinformatics systems must incorporate 'logics' that more resemble those described by Shackle, which are quite far removed from the logics discussed within the ambit of a conventional, logical-positivistically-orientated Artificial Intelligence. We shall return to the nature of these 'other kinds of logics' in Chapter 3.

It is instructive in this respect to review how we have over the last decade or so worked out some relatively successful strategies for modelling-technology transfers, strategies that are successful precisely because they are able to come to terms with the specific social forms of despair of the third world (*e.g.* in Bangladesh, India, and, albeit to a lesser degree, Thailand). Now the problem arises of extending these strategies in such a way that they promote the introduction of hydroinformatics in the third world. Let us, then, review some of these strategies in relation to hydroinformatics at a pragmatic level.

The first that has been followed in transferring modelling-technology strategy is that of maintaining a *stable working environment*, in the most immediate, material sense. This is one in which the reliability of equipment and ancillary services is assured to an acceptable level and where, correspondingly, the personal development of the staff has a perspective and a continuity that promotes morale. However, this development of the staff must then be seen within the context of a project in the third-world country which is supported both by first-world staff and, more significantly in this place, by first-world products. Indeed, the key feature in this strategy is one of a proper interplay between projects and products. So far in time, the products that have been most relevant in this connection have been third-generation modelling codes, but, increasingly, fourth-generation modelling is taking over this role and extending it further. What happens, in effect, is that the original first-world revelation that is encapsulated in the product codes comes to presence within the project team, within the third world. A new interplay then occurs between the local view of the physical prototype and the representation of that prototype that 'comes to presence' through the revelatory insight provided by the codes. The stability of

---

[7]It might be thought that the anthropological and more general social investigations that are now being introduced into water-resources projects in the third world should have their counterparts in water-resources projects in the first world. There are, however, such dangers to social stability that can follow from investigations of this nature in first-world societies, that this need is nearly always rejected, (I have gone into the reasons for this in some detail in my — unpublished — *magnum opus*). Thus in first-world projects, ecological investigations are only accepted in so far as they do not call social structures into question. Indeed, we were all in 1989, in particular, witnesses to the most extreme social-destabilising effects of such questionings in the case of second-world societies.

the project environment is thus based upon the authenticity of the products that enter the project, and so upon the stability of the overall modelling paradigm.

The second, closely-related strategy is that of maintaining a continuity in the working environment in a longer-term, more social-psychological sense. This is now seen to be based upon a continuity in the paradigm. It necessitates the taking of a very carefully-considered quantum jump in the level of the technology, so that the number of such jumps will be restricted while still allowing the project the widest possible scope for further development. It is upon this foundation that it becomes possible to build an environment in which technology comes to present its essential aspect, as a 'mode of revealing', so that 'technology comes to presence in the realm where revealing and unconcealment take place, where *alētheia*, truth, happens'. This is then an environment where not only 'truth appears', but where, as it appears, it is sustained with solicitude. It is this environment that instills a new hope into those who participate in the process of the revealing and sustaining. If we may be permitted to resort to a current cliché, we might describe this strategy as being one of the 'implanting of centres of excellence' in third-world societies, and situating these at the most critical points in the decision-triggering information networks. The challenge now must be to introduce hydroinformatics systems that perform a similar function. More fundamentally, the authenticity that is revealed in the modelling comes into synergy with those authentic traits that provide the most deeply-rooted cultural foundations of the society — so that this is a synergy that occurs already here at the deepest levels of the Jungian 'collective unconscious' (Jung 1943/1953), at what we might try to describe at this point already as 'the level of the religious'. In a much deeper sense again, 'the numbers have begun to function in another way'.

When seen against this background, the prospects for applying hydroinformatics in the third world are much brighter than they might at first sight appear. Experience with modelling-service-transfer projects suggests that it will be practical to transfer hydroinformatics system also into the third world, as one of the many paths that could ultimately lead to 'one world'. Certainly, hydroinformatics cannot be a party to any kind of 'technological apartheid'.

2.4 The European dimension of technology, and thus of hydroinformatics also

Hydroinformatics is the hydraulics of the informational revolution. But what is the informational revolution in its essence? The informational revolution is a technological revolution, which is to say that, in its essence, it is a revolution in the manner of revealing, in the nature of the coming-to-presence of 'the new'. What is it then, essentially, that is revealing itself, that is coming-to-presence, in the outer-garments of technology, both of industrial and informational revolution? This question was in fact posed distinctly, albeit in another form, already at the beginning of the last century, and a name was coined in order to answer this question in its nineteenth-century guise. This was the name of *nihilism*. As the last century progressed it became clear that this was a movement that was proceeding at the greatest depth in the collective human psyche, at 'the level of the religious'. For his part, Heidegger (1977) arrived at one of his most striking insights in a sequel to *The Question Concerning Technology*. He there took up Nietzsche's most famous (or, for those unacquainted with its context, most infamous) Section 125 of *Die Fröhliche Wissenschaft* (*La Gaya Scienza*) — that in which Nietzsche's Zarathustra proclaimed that 'God is dead, and we have killed him' — and saw the distillation of this section in one, later aphorism of Nietzsche:

> The greatest recent event — that 'God is dead', that the belief in the Christian God has become unbelievable — is already beginning to cast its first shadows over Europe.

Heidegger observed of this:

> From this sentence it is clear that Nietzsche's pronouncement concerning the death of God means the Christian God. But it is no less certain, and it is to be considered in advance, that the terms 'God' and 'Christian God' in Nietzsche's thinking are used to designate the suprasensory world in general. God is the name for the realm of ideas and ideals.

To Nietzsche's rhetorical question: 'Are we not straying as though through an infinite nothing?', Heidegger responded (pp. 60-61):

> The pronouncement 'God is dead' contains the confirmation that this Nothing is spreading out. 'Nothing' means here: absence of a suprasensory, obligatory world. Nihilism, 'the most uncanny of all guests', is standing at the door.

> ... Nihilism is a historical movement, and not just any view or doctrine advocated by someone or other. Nihilism moves history after the manner of a fundamental ongoing event that is scarcely recognised in the destiny of the Western peoples. Hence nihilism is also not simply one historical phenomenon among others — not simply one intellectual current that along with others, with

Christendom, with humanism, and with the Enlightenment —
that comes to the fore within Western history.

Nihilism, thought in its essence, is, rather, the fundamental
movement of history of the West. It shows such great profundity
that its unfolding can have nothing but world catastrophes as its
consequence. Nihilism is the world-historical movement of the
peoples of the Earth who have been drawn into the power realm
of the modern age.

In another place — in a work unfortunately unpublished — I have
bodied-out this insight into a complete characterisation of the informational
revolution. This revolution then appears as the outer-world name and form of
the surfacing, or coming-to-presence of nihilism, and indeed — as presaged also
by Nietzsche — it appears as a surfacing or a coming-to-presence of nihilism
in its most critical form, in the form that the younger Barth (1922/1933, *e.g.* p.
225) already associated with the word *Krisis*. Just as we may discern the focus
of nihilism (that is driven by what theologians, at least from Augustine to Barth,
associated with 'nothingness') within the depths of the collective unconscious of
the peoples of Europe, so we can discern the focus of the informational
revolution also within the darkest recesses of the collective unconscious of the
European peoples.[8] The crisis of and the confrontation with nihilism — which
is of the essence of the informational revolution — then has its focus within the
thought-world of Europe. It was in this same understanding that Heidegger
entitled one of his last works on this subject *Der Europäische Nihilismus*, while
Nietzsche spoke (1887/1969, p. 501), with his customary penetration, of
Schopenhauer's contribution to the overthrow of the theistic paradigm:

> ... der Niedergang des Glaubens an den christlichen Gott, der
> Sieg des wissenschaftlichen Atheismus, ist ein gesamt-

---

[8]As the informational revolution progresses, so an ever-increasing effort is put into processing
information, so that information processing comes to employ ever more persons, and indeed it
already employs more persons than are employed in agriculture and manufacturing combined in
many societies. Information, however, is *defined*, and certainly since Szilard, Shannon and
Brillouin, as that which resolves a state of uncertainty, so that an informational society is one that
is driven by an ever-mounting *sense of uncertainty*. This is then to say that, in its depths, it is
driven by anxiety, by despair and, ultimately, by dread. It correspondingly, on the one hand,
spends an ever-increasing amount of time and effort on distractions of every kind, such as
saturation television viewing, even as, on the other hand, it embarks upon renewed searches for
inner certainty. It thus follows from its very origin that any serious analysis of the informational
revolution must be based on the soundest theological, and so dogmatic-scientific, foundation
available.

europäisches Ereignis, an dem alle Rassen ihren Anteil von
Verdienst und Ehre haben sollen.[9]

Thus the most obvious 'world-catastrophic consequences' of nihilism —
the two world wars of our century and the destruction of the divinely-
establishing symbiosis between Christendom and Judaism that accompanied the
second of these — had their focus within Europe. As Barth observed, in his
celebrated commentary on Heidegger and Sartre (1960, p. 345):

> We experience nothingness, and in so doing we experience
> ourselves and all things as well. Heidegger's astonishment is no
> less eloquent of this than Sartre's defiance, nor does the latter
> bear lesser witness. Their thought is determined in and by real
> encounter with nothingness.

> Their thought and expression are determined in and by the
> considerable though not total upheaval of Western thought and
> expression occasioned by the world wars.

> In our time man has encountered nothingness in such a way as
> to be offered an exceptional opportunity in this respect.

In Barth's later works, the realm of an all-destructive nihilism becomes
translated into English as 'the kingdom of nothingness'. In his *Church
Dogmatics*, Barth explained how that motion in the depths of the collective
unconscious that is the innermost expression of nihilism has continued to take
on ever more concrete names and forms in our outer world. Today we see in
the destruction of the natural environment, in the undoing of the creation, the
further allegorical representation of nihilism, and one which gives expression
to 'the kingdom of nothingness' in the most concrete and material way possible.

Now of course the manifestations of an ever-encroaching nihilism is by
no means a feature that is peculiar to the European peoples. By no means is it
this! It is however *centred upon Europe*; it has a focus, in the same way as does
a physical sickness, and this focus is situated in the collective psyche of the
European peoples. Then, in so far as hydroinformatics constitutes in its essence
one part of the total field of contest of nihilism, so it follows that *hydroinformatics begins as a European possibility*.

---

[9]*Author's translation:* The decline of belief in the Christian God, the victory of scientific
atheism — this is a pan-European event in which all races will have their share of merit and
honour.

2.5   The market realisation of hydroinformatics

Hydroinformatics appears as an integral part of a technological revolution, and so a revolution in revealing, in the surfacing of the new into the already-established world, of the coming-to-presence of this new within an already-established world. Hydroinformatics thus constitutes one particular area in which the overall assault of nothingness and the confrontation with nothingness occurs. (In Nietzsche's terminology: of the transformation from 'incomplete' to 'complete' nihilism.) Hydroinformatics is the outer world expression, or allegory, of the movement of the collective unconscious that is behind this assault and its confrontation, as this is played out in our relation to the waters of the Earth, and so in our relation to the very arteries and veins of the Biosphere.

Hydroinformatics begins as a European possibility, so that the problem of the realisation of hydroinformatics is in the first place a European problem, which is to say that it is in the first place a European responsibility. The working-out of hydroinformatics — its realisation in our outer world of name and form — is then a problem and a responsibility in the first place of European hydraulics, and so of European hydraulicians. This working-out, this realisation, must then be seen against the background of other changes occurring in Europe, and the most immediate and striking of these changes are, of course, those of a political-economic nature.

It is in the nature of a transformation blessed by grace that it is supported by fortunate accidents. Correspondingly, it is the mark of a movement blessed by grace that it succeeds, so to speak, in 'planning its own accidents'. The transformations occurring at the political-economic level in European society then provide hydroinformatics with an entirely remarkable and indeed unique opportunity, in that the field of operations of hydroinformatics now attains to a united-European scale. This is to say that the paradigms developed for the working-out and realisation of hydroinformatics are able to attain to quite new, European proportions, as the quantitative expression of their European political-economic dimension. The immediate future development of hydroinformatics then corresponds to the playing-out of a transformation in technology within what is increasingly called 'the Europe of the inner market'. This is then a very large market indeed: taking water-related industries in the broadest sense, it supports an industry with a turnover of many tens of billions of dollars a year.

The European inner, or single, market is, then, the *opportunity* of hydroinformatics, it is its 'happy accident'. It is this which enables the paradigms of hydroinformatics to attain to a much fuller development than would ever have been possible had they been confined to the frontiers of the European nation-states. And we are, indeed, already witnesses in this respect to an accelerating restructuring of our outer world of knowledge industries in Europe: of new association, alliances, mergers, reconstructions, and business failures. What we observe, in fact, is a transformation at the European scale that appears to be driven by 'business enterprise'. We shall return to this in the next chapter.

It will be already clear, however, that business survival in 'The Europe of 1992' will depend in the water industry (of water authorities and associations in much of Europe, contractors, consultants, technological-service institutes and, even, university departments: the whole industry in fact) upon two, inseparable elements: the most suitable people and the most suitable tools. The first is certainly nothing very new, even if it is now more strongly emphasised and introduces new social features, but the rapidly increasing emphasis on relevant and effective tools, and properly-integrated tool sets, is something quite new in the field of hydraulic engineering and water-resources management. As our millennium comes to an end, survival in knowledge industries generally will increasingly and ever more decisively depend upon the assembly of the right tools, and the effective integration of these. The problems of productivity, of overall effectiveness, of competitiveness and, in the last analysis, of *authenticity*, become dominant problems in all branches of knowledge industries. All of these problems call for creative thinking, but now this creative thinking acquires a quite other span within which to stretch its wings.

As there is here a risk of misunderstanding, some further clarification is necessary on this point. By this European dimension is to be understood only 'an opportunity to fulfil an obligation'. There can be no presumption of a European precedence in any other sense than this, and those of us in Europe must never claim anything more. Certainly the last thing this world needs is another bout of Euromania! The realisation of hydroinformatics is simply a *duty* that European hydraulicians *owe* to the world.[10]

Let us, following upon this, return to the influence of business, but now following Heidegger (1977, p. 64):

> Into the position of the vanished authority of God and of the teaching office of the church, steps the authority of conscience, obtrudes the authority of reason...

> The flight from the world into the suprasensory is replaced by historical progress. The other-worldly goal of everlasting bliss

---

[10]There is something ironic that I write this when European industry generally is becoming less and less competitive *vis-à-vis* Pacific-rim economies. However, we are here talking about a *responsibility* that has to be accepted in the first place by European industry, and on the whole this is being accepted, and its obligations are being met. Of course, what happens later, as industries from other societies take up hydroinformatics, is another matter.

is transformed into the earthly happiness of the greatest number. The careful maintenance of the cult of religion is relaxed through enthusiasm for the creating of a culture or the spreading of civilisation. Creativity, previously the unique property of the Biblical God, becomes the distinctive mark of human activity. Human creativity finally passes over into business enterprise.

Within the 'Europe of 1992' and indeed — who knows? — perhaps within an even 'greater-Europe' also, the social transformation that we associate with a technological revolution, the informational revolution, is one that is outwardly driven by 'business enterprise'. In this form it acquires all the attributes of the Heideggerian 'mundane and average world': it also has to do with competition and collaboration, planning and opportunism, marketing and selling. This is necessarily its way of proceeding within its social environment. Let us then consider how hydroinformatics is realised within this environment of 'business activity'.

When seen within most contexts outside of the USA, we can conveniently divide-up the knowledge industries that operate in water resources into two kinds: consulting and contracting. Within this view, we may see the essential function of consulting as one of *coping with complexity*. It is the primary task of the consultant to project and to control the realisation of the new which is coming-to-presence in the world: the construction, the management system, the on-line control system, etc. In effect, the consultant makes a business out of the problems of complexity, by reducing the complexity experienced by his client. The task of the contractor in this view, on the other hand, is essentially one of *organisation* and *integration*. It is the primary task of the contractor to bring-to-presence in the world that which the consultant projects and controls in its coming-to-presence. The activities of the consultant and the contractor can also be seen, in this light, as 'orthogonal'. It follows from this that the consultant and contractor may be expected to place different kinds of demands upon hydroinformatics, so that these demands may very well again proceed, analogously, in two, mutually-orthogonal directions. This situation is modified in the USA to the extent that Federal and State agencies may appropriate the roles that both consultant and contractor fill elsewhere.

Now within the scheme of generations of modelling, and so from the viewpoint of hydroinformatics as an outcome of numerical simulation modelling, the hydroinformatics system is a fifth-generation system. This is to say that it is there viewed as a tool, language or environment. It is then to be expected that one kind of hydroinformatics system may well be used by consultants and quite another kind again by contractors.

It has already been remarked that, on the side of the consultants, business survival in the 1990s will depend upon the right choice and combination of tools relative to available human resources. Appropriate tooling is the *conditio sine qua non* for business survival. The engineer working in this area will then have to transform, over only a few years, from one who uses fewer tools than almost any other professional person to one who is a tool-user on quite a large scale. These tools will take over to a large extent the 'lower-level' functions of the consulting engineer, leaving this engineer with the 'higher functions'.

From this it in no way follows, however, that the number of persons employed in consulting activities will decrease. It is in the very nature of the informational revolution that it cannot allow this. Experience with advanced numerical modelling alone shows that the proper use of tools in this area is exceedingly high-level-labour intensive, involving far more extensive field investigations, parameter identifications and determinations, multiscenario runnings, calibrations, validations and evaluations. The realisation of hydroinformatics will raise this level of labour intensiveness yet further again. By these means it is to be expected that the number of persons employed in knowledge industries on the consulting side will increase markedly under the influence of hydroinformatics. The greater costs involved will, however, be recovered from the even greater savings that can thereby be made in applications, and especially so when these applications have as their aim the avoidance of environmental catastrophes. In this respect, not so very much has changed over the last thirty-odd years (Abbott, 1959).

The assessment of the impact of hydroinformatics on the contractor is another matter again, and one that so far does not appear to have been much investigated. When seen from the point of view of productivity and competitiveness, the essential concept in fifth-generation modelling is that of 'the tool'. It is through the use of tools that productivity is increased most markedly, and quite commonly by orders of magnitude. The hierarchical superposition of tools provides additional leverage again, whereby gains in productivity can become nothing less than prodigious. This potential of tools and tool-sets has been realised most dramatically in manufacturing industries: in one glass plant, for example, one person alone 'produces' 1,800,000 bottles a day, and even of a variety of different shapes and sizes! Considerable gains are nowadays also realised quite widely in agriculture. When seen in this perspective, however, contracting remains one of the last 'islands' of low-level-labour intensive industrial activity. It is, correspondingly, an area where the most is to be gained in productivity and competitiveness by appropriate investments in tooling.

Now when we compare consulting and contracting in this light we observe that, whereas hydroinformatics has so far advanced along the axis of consulting primarily through its coming-to-presence from out of simulation modelling, its advance along the 'orthogonal' axis of contracting appears to proceed primarily through its coming-to-presence from out of data-base management systems, and so from out of data modelling. The link becomes one of project overviewing and simulation modelling on the one side and project organisation and data-modelling on the other side. The successful contractor of the 1990s will have to accumulate a great variety of high-technology construction equipment at his site, but the individual items of this 'machine park' will be required only at quite specific periods within the work. The relation of his plant availability to material supplies, labour availability, weather conditions and all other such ancillary factors, is one that calls for a wide range of hydroinformatics products. The aim of the contractor in a competitive situation is to use machinery as intensively as possible, to process material as quickly as possible (thus reducing to a minimum the time between his payment for equipment and materials and his being paid by his client for the completed work), and to utilise his labour force to the best advantage. The principal aim of hydroinformatics is, then, to assist in realising these aims, always under the special conditions of uncertainty that are inherent in the natural and social environment of the contractor.

We observe further, in this respect, that a marked shift in priorities in all such engineering activities is now becoming apparent. Whereas throughout most of the 1980s the first priority was quality, towards the end of the 1980s Japanese manufacturers evolved another ranking, and one which is receiving increasingly widespread recognition (*e.g.* Lorentz, 1989). This reduces quality to a 'qualifying factor' only — one that is essential just to stay in business at all, placing it fourth in order after speed of reaction to customer requirements, direct labour productivity and turnover of work in progress. These priorities are then seen to correspond rather well to the benefits to be expected from the use of hydroinformatics systems, certainly in contracting, but also in consulting.

There are, of course other agents than the consultant and the contractor involved in the process of the bringing-to-presence of the new into the existing world, even though the consultant and the contractor are the main representatives within the traditional knowledge industries. The most significant of the other agents is one that is not always considered within the area of knowledge industries. We may perhaps best describe it as that comprised of 'the legal' and 'the political'. Although this agent is itself driven by such social influences as the environmental movement, it is the legal and the political entities which give a definite form to the aims of such movements within the context of society as a whole, by negotiating contracts, by promulgating acts of law and, further, by *institutionalising* the corresponding social arrangements. Moreover, it is this agent that places some of the heaviest burdens upon society as a whole, by forcing changes in farming practices, in all manner of water treatments, in industrial processes, and in other ways.

Just as we may make a certain tenuous link between consulting and simulation or process modelling in the integrating direction and contracting and data-base management or data modelling in the organising direction, so we may now perhaps discern a primary link between the legal-political and artificial intelligence in the institutional direction. Hydroinformatics will have to provide the tools upon which legal and political decisions are both based and implemented, and when seen from this point of view the direction of advance of hydroinformatics would appear to proceed predominantly along the axis of those natural-language-manipulating codes that simulate intelligent behaviour while using numerical models and data bases as encapsulations and repositories of a 'deeper knowledge'. Hydroinformatics then appears as the principal means to navigate the treacherous political waters between macro- and micro-business economic decline and environmental deterioration, with all its legislative consequences. This situation would appear to obtain within the U.S. environment as well as the European, although it is often rather less clear there due to the special position of U.S. agencies. For example, already in 1979 direct and indirect U.S. Federal expenditures for model development and dissemination passed through the 50 million dollar level, while Jamieson states (see De Coursey, 1990, p. 706) that the Soil Conservation Service of the U.S. Environmental Protection Agency alone is spending 150 million dollars on Geographic Information Systems over a current five-year period.

The consultant, the contractor and the legal-political agents provide the primary market for hydroinformatics, so that they are the principal means for realising hydroinformatics on its market side. This is then to say also, however, that the three axes of advance of hydroinformatics at the level of abstraction are mirrored, even if only weakly and tenuously, in the three axes of advance of hydroinformatics at its most gross level, at the level of its market for applications. Quite symmetrically, these business aspects of hydroinformatics, that are predominant in the present stage of development of the European dimension, leave their mark already on the designs of the systems themselves. For example, the architectures of hydroinformatics systems must reflect the ways of working of the organisations within which they are installed, and indeed it increasingly appears that system architectures must be increasingly influenced by business considerations.

There is another way of describing the changes in institutional structure concomitant with the single, inner-European market. This proceeds from the thesis that a large number of organisations working in the field covered by hydroinformatics and its applications — the majority in fact — must now pass from the non-traded or 'sheltered' part of the economy to the traded, or 'unsheltered', part. This is to say that these organisations must pass from a present situation, where they are scarcely subjected to market forces at all, to one where they must face competition, and indeed increasingly intense competition, over an increasingly wide range of their activities.

For example, many institutes have been established on a national basis with state funding for local applications. They have long lived a modest but comfortable existence on the basis of an ongoing study of national projects. There has accordingly been little incentive for these organisations to innovate, to enter foreign markets and, indeed, to acquire any commercial competitive potential at all. With the advance of the single market, this situation changes drastically. This advance exposes organisations with inadequate leadership, unsuitable management, insufficient and ill-directed investment and undeveloped marketing capabilities to the effects of competition from organisations that are better prepared in this respect. Given their starting positions, it must be exceedingly difficult for these organisations to change direction and to restructure in time to survive. In any event, their survival depends upon their acquiring appropriate tools and tool sets, but even then many of these organisations lack the capabilities to select and integrate tools into tool sets, so that tooling, in and by itself, will not save them. Some will no doubt survive by changing their nature, becoming, so to say, 'local tool-using specialists', but others will just as surely be surpassed by their own former clients, so that even this kind of niche becomes closed to them.

This expected loss of organisations is not, in itself, so very unfortunate, but the question naturally arises of the effect of all this on professional employment. It has already been explained that hydroinformatics will have the effect of increasing employment markedly in those areas in which it is applied. On the other hand, as the number of individual organisations is reduced or their structure is changed, so a large number of professionals will need to make career adjustments. These adjustments will become more difficult, however, precisely because the industry that hydroinformatics serves is moving increasingly from the non-traded, sheltered part of the economy to the traded, unsheltered part. Previously the response of any traded-economy-orientated organisation in a business turn-down was to advise and assist the transfer of their excess people to the non-traded-economy organisations, where requirements were generally less demanding. It is by no means certain that this response will remain viable within the situation that is now appearing. It follows that the organisations that have only newly entered the traded economy will have to invest in upgrading their staff, suiting them to new opportunities, if they are to avoid unemployment, with its debilitating effects on morale.

Already here we see how a change in institutional structure is associated with a change in what has been called, following J-J. Rousseau, the 'social contract'. In the informational revolution generally, there occurs a *transformation in the social contract*, and this has many other manifestations. One of the most striking of these is that in which persons who have traditionally thought of themselves as 'pure' scientists (*e.g.* most biologists and ecologists, many if not most hydrologists) are rather suddenly called upon to 'serve society' as 'applied' scientists. Such persons are, however, intrinsically *incorrigible hobbyists*: they are individuals who follow an own bent, for whom a study is properly made only 'for its own sake' — for whom, for the most part, the very notion of 'business enterprise' is anathema.

Now of course, on the one side, society has to be very glad that there are such incorrigible hobbyists, for otherwise there would scarcely be any biology, zoology, ecology, or other such subjects. Certainly civil society has hardly ever deemed it necessary to encourage the respective practitioners by offering them substantial material rewards. (As one of my colleagues observed: 'Ecologists are as poor as church mice'!) But as a result, and on the other side, many such scientists can feel no obligation whatsoever to participate in hydroinformatics exercises directed in a 'business-like way' to social-economic ends. We have in fact here to do with a very deep-seated problem — one as deep as nihilism itself — and one which hydroinformatics cannot ignore. In effect: *hydroinformatics must accommodate itself to hobbyism*. As one means of accommodation, it is to be expected that precisely to the extent that hydroinformatics concerns itself with the intelligent behaviour of organisms, even if only through their representations as intelligent agents, so it will encounter rather more sympathy from biologists, zoologists and others who experience a natural empathy with the subjects of their studies.

This finally brings us to the question concerning the relation of hydroinformatics to hydraulics. Why, after all, should the main responsibility for the elaboration of hydroinformatics systems, with all their biochemical, ecological, anthropological and other such aspects, devolve upon the *hydraulician*, of all people? The answer should now, however, be clear: it must do so because the *hydraulician is the best prepared of all people to take up this challenge*. The hydraulician is the *technologist* with the mathematical bent, he who bears the heritance of Leonardo da Vinci, Galileo, Newton and Euler, he who knows most surely that 'pure thirst is only quenched in pure water'. It is the hydraulician who has most prepared the social ground for this great venture, by advancing the generations of his models and by establishing the corresponding business and other institutional structures on a world scale. It is the hydraulician who follows business trends and tendencies, such as those of 'global localisation', and who is now making his own arrangements for conglomerate and transnational organisations. The task thus falls to the hydraulician, even if only apparently by default.

3 The logics of hydroinformatics

I think I have at last made you realise one thing, Aristos, that any expression of an abstract idea can only be an analogy. By an odd fate, the very metaphysicians who think to escape the world of appearances are constrained to live perpetually in allegory. A sorry lot of poets, they dim the colours of the ancient fables, and they are themselves but gatherers of fables. They produce white mythology.

Anatole France, *The Garden of Epicurus*

3.1 The logic of hydroinformatics in the positivistic and empirical sense and in the sense of dogmatic science

Hydraulics is important! Just as the waters are among the most precious of all that we have here on Earth, so the study of the motions of the waters and all that these motions provide is among the most important of all studies. This simple fact has long been overlooked, so that for a long time now the subject of hydraulics has not been regarded with the esteem and respect that properly accrues to it. Indeed, even to the extent that anyone says what I have just now said, so they stand today accused of exaggeration. For this situation, however, the hydraulicians have largely themselves to blame, for they have done little to espouse the cause of hydraulics in society at large, and have themselves been only too content to allow their subject to assume an ever more lowly status within the social scale or ranking of the technologies. For far too long now, hydraulics has assumed the cloak of the moderately useful, but rather uninteresting. Whether assumed on the grounds of the modesty or the timidity of the hydraulicians, however, this cloak has done no good to hydraulics, while its assumption does an ever-increasing harm to the world that hydraulics has to serve. It is now high time for hydraulicians to stand up and insist that *hydraulics is important!*

Now it is no use saying that a subject is important if one cannot show how it can be applied within our current societies. This, however, is the very function of hydroinformatics. For ours is an informational society, where a subject attains to social significance to the extent that it can be 'informationalised'. As we have now seen, hydroinformatics is the specific mode of coming-to-presence of hydraulics that must obtain within an informational society. The development of hydroinformatics is, similarly, an inseparable part of the development of the informational revolution generally. It is then subject to the same laws as the informational revolution generally, as has been argued earlier. As I have tried to explain in this text, following Barth and Heidegger, hydroinformatics must be seen within the context of technology generally, as one of the fields where the confrontation with nihilism, with nothingness, is at its most intense. As such, hydroinformatics cannot be separated from other elements of this confrontation, such as the 'Green' movement, the move to environmental protection generally, to the conservation movement, and the overall tendency towards an individual and a social solicitude towards the entire creation. I have tried to show how hydroinformatics interacts with these movements, reinforcing further the process of political action, legislation and decision-making generally. Obviously such interventions as these cannot be realised without some insight into their philosophical and theological background, and I have not hesitated to introduce these background elements. The problem is thereby posed, however, of the manner in which hydroinformatics should consider such philosophical and theological matters.

The response to the very posing of this problem is, of course, entirely predictable. Those with a penchant for such matters will be enthusiastic about a rather thorough treatment, while those for whom such matters are not attractive will dismiss them as entirely irrelevant. Moreover, these attitudes will have nothing much to do with scientific attainment, for science most commonly thinks itself far above these matters. It is indeed a most persistent and remarkable feature of the human mind that when it encounters something that is far above it, it commonly imagines that this is far below its own level. So it must also go with any discussion of metaphysical and dogmatic-theological aspects.[11]

---

[11]Probably no author's work has suffered so much in this way as that of Kierkegaard, who was for Wittgenstein, no less, 'by far the most profound thinker of the last century'. The process started already in Kierkegaard's own life time with the vituperations of Aaron Goldschmidt, and continued with Georg Brandes' gross 'popularisation' of Kierkegaard, of 1877 — which however did more than anything else to promote the dissemination of the works themselves in other languages. The business of trivialising, denigrating and deriding Kierkegaard has continued into our own time (*e.g.* Wahl, 1959/1969).

Of course, if one just sees the whole matter from the side of the 'dismal science', of the ordered, the numbered and the calculated, it is all very easy. In this case one is back with logical positivism and logical empiricism, and one can only let one's authors run downwards, from the narrow brightness of Carnap (*e.g.* 1937, 1950, 1956) to the general dullness of Popper (*e.g.* 1936/1959; 1963, especially pp. 253-292: see in this respect, and with reservations, Habermas, 1968/1972). But enough has been written above to indicate that taking this approach, as and by itself, would be disastrous. There are, however, here already points of reference to Husserl and others of the phenomenological movement. But then, if we should follow these up in the spirit of modern technology, we are bound to find ourselves moving rapidly away from logical-positivistic and empirical thought, towards 'hermeneutic' phenomenology, towards Heidegger, where (1977, p. 112) 'thinking begins only when we have come to know that reason, glorified for centuries, is the most stiff-necked adversary of thought'. Such remarks — even though common currency since the Vedic masters of some three thousand years ago — will still be anathema to most contemporary scientists and engineers. If we are then to go beyond this again, to relativise it, such as by exercising the *credo ut intelligam* of the Christian Church, then we will be well and truly out on a limb! But if, on the other hand, we do not do this (Barth, 1960, p. 378), 'if we are merely clever and are obviously not prepared to practise the *credo ut intelligam*, we are well advised not to enter into this sphere, or of dogmatics generally. For we can have no prospect of success.'

The problem really then comes down to this: that one cannot even begin to talk about the informational revolution and all that is entrained with it, without resorting to dogmatic categories, even though this approach is highly uncongenial to the usual ways of thinking of modern science.[12]

---

[12]The necessity of such a dogmatic foundation for technology was an essential insight of Heidegger, even as Heidegger saw the relation obtaining here as one proceeding in both directions, in that he simultaneously developed the ontological ground of metaphysics through a 'technological' vocabulary and overall approach. But then, as Barth (1960) so perspicaciously observed, there is something highly irregular in Heidegger's most basic ontological concepts of Being and nothingness. In my — most unfortunately unpublished — *magnum opus*, I have followed up this fault that runs through Heidegger, with its always dangerous tendency to the 'occult' distortion of religious experience. Heidegger always tends towards a position, where comprehension dominates over faith and the *credo ut intelligam* is not practised in proper equilibrium. (Thus, as I have explained in some detail in my unpublished principal work, it was just as inevitable that Heidegger should have sympathised with the 'black mythology' of Nazism as it was that Barth should have been one of Nazism's most implacable opponents). Accordingly, even while Heidegger's works are required reading for all strategic thinking in technology — and the 'higher' the technology, the more required — it is necessary to be aware of the basic nihilistic fault line that runs through all of these works.

Nothing could illustrate these matter more clearly (or, for those who think in positivistic terms, nothing could obfuscate them so completely!) as the thesis that has been so provocatively introduced here, that 'the numbers have begun to function in another way'. Let us first go over this from the side of numerical modelling, but now stepping back from the immediate picture, and even from the entire picture that unfolds in our own times. For, so long as we take the shorter view, we must consider numerical modelling as an activity with scientific, æsthetic, business, legal and other such attributes of the Heideggerian 'mundane and average world.' Indeed, if we stand close enough, we might even be persuaded that all that is required from numerical modelling is a small set of numbers or even a single number: Will the water level be 1.7 m or 1.8 m? Will the coliform count be above or below a certain threshold?; and so on. However, above and beyond all such immediate considerations we must at some stage pose a question of *purpose*, and of *aim*. What is numerical modelling *really* about? What is it that we are trying to do, *essentially*? We then have to do with a *teleological question*, which I have called the *teleological question concerning numerical modelling* (Abbott, 1989).

As has now been repeatedly emphasised in this work, whenever we pose such a question we have to be clear that this cannot be a purely scientific question. In fact, in purely scientific terms, this must appear as a quite meaningless question. Indeed in our Western tradition, and certainly since the time and works of Voltaire, teleological questions also have been relegated almost exclusively to the concerns of dogmatic theology. Moreover, just to make things more difficult again and as also introduced above, the theistic allegory itself has long been seen by scientists as an obstacle to scientific thought, with its very language antagonistic to the requirements of modern-scientific clarity. Thus as soon as we try to pose such a teleological question more precisely we must appear to move far away from the paradigm-systems and language structures of modern science. However, despite these difficulties, there seems to be no other way of posing the question than within the framework of the theistic allegory, and so we must proceed in this direction. The question then takes the following general form:

> *What is the role of the numbers in maintaining the covenant between the Creator and His creature?*

When posed in this form, two aspects of the question are at once thrown into relief. The first of these is that the question is older than recorded history, while the answer that was given to it appears to have remained fairly constant up until the dawn of modern science, since which time it has changed markedly (Abbott and Basco, 1989, pp. 367-373). In all societies previous to those initiated by the European Enlightenment, the numbers entered into the covenant through the agency of *number symbolism*. Correspondingly, it is one of the hallmarks of modern science that it has *completely demythologised the numbers*. In modern science, the numbers have been stripped completely of their allegorical attributes and correspondingly this process of demythologisation has been essential to the success of modern science. Today, the number symbolism that provided the very bearing structure of the thought worlds of the alchemists and the astrologers has been almost completely eradicated.[13]

However, it is well known (from psychology, anthropology, hermeneutics, rhetoric and any number of other such disciplines) that myths never disappear: our basic innermost psychic processes continue to operate and they must still find their expression in terms of our outer world of name and form. It has then been suggested (Abbott and Basco, 1989) that the modern form of number allegory is the numerical model. The numerical model is the name and form in which the numbers nowadays enter into and support the covenant. Now it is a basic lesson of modern dogmatic theology (*e.g.* Barth, 1960) that the image of the inexpressible, which is the allegory, takes increasingly concrete and physical-material forms. Thus, as has already been mentioned, Barth demonstrated convincingly how the world wars of the first part of our century were allegorical representations of that inner division of the psyche to which the last century attached the name of nihilism, and which we associate, to use the standard English translations of Barth's expression, with 'the kingdom of nothingness'. As has also been emphasised earlier, we see today in the destruction of the natural environment, in the undoing of the creation, the further allegorical representation of nihilism, and the one which gives

---

[13] Another explanation, another footnote! We may view astrology as a methodology for forcing the unconscious (understood in the very broadest possible sense) to 'give up its secrets' to the conscious (of the astrologer). Astrology used a correspondence between what we might nowadays describe as the DNA-coded perfection of our psychic apparatus and the only outer-world allegory for such perfection, namely the motions of the planets across the starry skies. As astrology progressed, so all entities within the world, whether mineral, vegetable, animal *or divine* came to be integrated into this system: astrology became totally holistic. As will be outlined further in the appendix to this chapter, the numbers enter into our thoughts as the ultimate 'ordering agents'. They were accordingly essential to the astrological *system* in its entirety, and so necessarily became imbued with spiritual, holy, or sacred powers: they provided an *essential* support to society in its intercourse with its gods (see Jung, 1952/1955). In the alchemical movement, on the other hand, the allegorical medium became a branch of mineral chemistry which is even today situated at a 'divide' in our scientific metaphorical structures, this being one that proceeds roughly along the line where reductionist reasoning (leading back to quantum theory) meets evolutionist reasoning (leading forward to ecology). The ancillary-religious nature of the practice of alchemy was recognised already by Hitchcock (1857/1977) and was finally brought out most clearly and emphatically by Jung (1944/1953) in our own century. Here again, the numbers entered as the ultimate ordering agents that supported the relation between man and what had become, within both the Arabic and the European contexts of that time, man's one and only God.

expression, in the most concrete and material sense possible, to the 'kingdom of nothingness'.

If we now associate these two lines of thought we see that numerical modelling attains to another kind of significance in particular in its relation to the environment, both in its first-world and third-world senses, and thereby it attains to a higher aim or purpose. We see that, to extrapolate the Barthian vocabulary, 'numerical modelling is the numbers' way of supporting and arming us in our struggle with the kingdom of nothingness'.

It is only against this background that we can really see through the playing out of the dramas and comedies of numerical modelling in this area: the competition and the cooperation, the strategic planning and the opportunism, the science and the commerce, the thought and the action, and indeed all else. However, before considering how this support is provided within the context of hydroinformatics systems which integrate both logical and numerical features, let us look further at the 'question concerning logics' and its formulation within AI.

3.2   α-logics and ß-logics

At one of these gatherings that pass by the name of 'receptions', I was introduced to one of Her Netherlands Majesty's ambassadors. He immediately explained — I think more apologetically than condescendingly — that he was an 'α-person' (*een α-mens*). He thereby intimated in the manner most appropriate to a professional diplomat that we could unfortunately have nothing further to say to one another, this following at once from my being introduced as an engineer, and so a 'ß-person'. Thus, solely through a form of introduction, a division was established in his mind between us, a division that was almost total, and for all practical purposes unbridgeable.

The nomenclature itself goes back to the division within the old Netherlands *gymnasia*, where the 'α-direction' was that of the classics, including the classical languages, while the 'ß-direction' was that of science, which was the direction that led to an engineering education. The division remains as strong and clear to this day, although younger people now tend to speak more of *A-people* and *B-people*.

Now what can it *be* that demarcates all persons with a higher education into just two such distinct classes in this way? Evidently there is a *world of difference* between them, so that, following Husserl, this difference must be grounded upon a *difference in logics*. The logical processes of the 'α-mind' must then be *essentially different* from the logical processes of the 'ß-mind' and this difference in logics must have been established already at the most impressionable of ages. We may, for our immediate purposes, call the logic of the scientific mind — which will most probably be that of the reader of this work — the 'ß-logic', and the logic of the α, or classical mind, the 'α-logic'. Of course every logic itself implies a science. We shall then follow the European tradition, the tradition of Christendom, and view the α-logic, in its most pure and refined form, as the logic of dogmatic science. Within the frame of reference of a theistic religion, and most emphatically within Christianity, it is this α-logic that underlies the whole science of dogmatic theology. We thus come to see this division as one that is grounded at the level of logic itself, as a difference between the logic of modern natural or physical science and the logic of dogmatic science. We then recall, following Nietzsche, that this difference was promoted every bit as much by the men of the cloth as it was by the men of natural and physical science. Today, we see the ß-logic expressed most directly in the works of Frege, Carnap and a whole torrent of productions of computer science, and especially of artificial intelligence. We similarly see the α-logic most clearly in the world of modern art generally, in the surrealist movement in its broadest sense, in modern hermeneutics and rhetoric, and in existentialist philosophy. There does indeed appear to be very little in common between these, even though the one seems to be as necessary to us as the other. Most readers will need little persuasion about the need for a 'ß-logic'. Concerning 'α-logics', however, we may follow Barth, who expressed the matter as follows in his commentary on Heidegger and Sartre (1960, p. 345):

> We may certainly learn from them, if we have not learned it already, a more intense and acute awareness. In this sense, whether taught by Heidegger and Sartre or elsewhere, no one today can think or say anything of value without being an 'existentialist' and thinking and speaking as such, *i.e.* without being confronted and affected by the disclosure of the presence and operation of nothingness as effected with particular impress-iveness in our day. Whoever is ignorant of the shock experienced and attested by Heidegger and Sartre is surely incapable of thinking and speaking as a modern man and unable to make himself understood by his contemporaries.

What now is the main characteristic of this α-logic, of the kind that we shall have to introduce into our hydroinformatic system environments as soon as we try to come to grips with the social-anthropological problems of water-resources problems in the third world, and which indeed already intrudes as soon as we proceed at all far in the ecological dimension of hydroinformatics? Barth (1960, p. 293-294) spoke in this respect of a necessary 'brokenness' of all theological thought and utterance. He explicated this as follows:

There is no theological sphere where this is not noticeable. All theology is *theologia viatorum*. It can never satisfy the natural aspiration of human thought and utterance for completeness and compactness. It does not exhibit its object but can only indicate it, and in so doing it owes the truth to the self-witness of the theme and not to its own resources. It is broken thought and utterance to the extent that it can progress only in isolated thoughts and statements directed at different angles to the one object. It can never form a system, comprehending and as it were 'seizing' the object.

... But why is this true ... universally? The reason is obvious. The existence, presence and operation of nothingness are not only the frontier which belongs to the nature of this relationship on both sides and which is grounded in the goodness of the Creator and that of the creature. They are also the break which runs counter to the nature of this relationship, which is compatible with neither the goodness of the Creator nor that of the creature and which cannot be derived from either side but can only be regarded as hostility in relation to both.

... (Thus) not even objectively is the relationship between Creator and creature a system. It is always broken by this alien element. ... Here, if anywhere, it is imperative that theology, which is also a creaturely activity, should acknowledge that it is bound up with nothingness, and cannot and must not try to escape it. Here if anywhere theology as the subjective reproduction of objective reality ought not to impose or simulate a system. Here especially theology must set an example for its procedure generally, corresponding to its object in broken thoughts and utterances.

Technology always asks 'how?', while science always asks 'why?' Science answers its questions through its logics, α-science through its α-logic and ß-science through its ß-logic. Technology answers its questions by referring to both indifferently, although with an emphasis on the α side in some situations, and on the ß side in others. Our hydroinformatics systems, as products of technology, have to comprehend both: they must proceed in a ß-logic, as already well rehearsed in computer science generally and in artificial intelligence in particular, and, it now appears, they must attain the capacity to proceed in an α-logic as well. We are then led to ask: is this last requirement an impossibility within any world of first-order languages — perhaps even the very impossibility can be demonstrated? — or is it still possible, using a 'device' that, in its turn, 'corresponds to its object in broken thoughts and utterances'? Such a device must clearly work with signs of a finer grain than the alphabetical and other symbols of a ß-logic, so that it must be in some way *sub-symbolic*. But then we see at once, of course, that this is precisely the feature which is most strongly promoted by the partisans of connectivism, who are the partisans of the subsymbolic paradigm, this being understood as something that can be realised using neural-network devices (*e.g.* Rumelhart *et al*, 1987/1988). However, one meets other persons working in AI than connectivists who appear to be groping towards a similar objective. The one discovers that contra-trend judgements in business decision-making (*i.e.* individual or group judgements which differ sharply from the trend of opinion within the company or industry) cannot be treated by such methods as protocol analysis (that appears to derive from Carnap), even as it is just *these* judgements that provide the most powerful competitive potential. Another again seeks only a better foundation for a truth maintenance system, but, in his search for a more secure 'deep knowledge', discovers that the external representation as information of this knowledge makes quantum semantic leaps. Also in this, for ß-people most strange and unfamiliar of all strange and unfamiliar places, one meets others who are groping their way around. For those of us here, the conceptual instruments of ß-science, and indeed its entire ß-logic, no longer suffice, and we begin to take up the conceptual instruments of an α-science, with its α-logic of 'broken thoughts and utterance'. (And then we are naturally led to the outermost limits of even these instruments, to theology and its dogmatic science).[14]

However — and this is now a *very* crucial '*however*' — our hydroinformatics systems are not purely logical systems, but they are also and inseparably numerical, or algorithmic, systems. As soon as we pose our teleological question concerning numerical modelling within the context of hydroinformatics systems, so we have to ask how the numbers lend their support to the covenant within a logical context that is, in its turn, one of both α-logics

---

[14]Neural machines are especially suited to establishing transformation processes with a minimum of preconception. They thus correspond to the principal requirement of anthropology set out by Lévi-Strauss already in 1962 (p. 117), as follows:

Plus nos connaissances s'accumulent, plus le schéma d'ensemble s'obscurcit, parce que les dimensions se multiplient, et que l'accroissement des axes de référence au delà d'un certain seuil paralyse les méthodes intuitives: on ne parvient plus à imaginer un système, dès que sa représentation exige un continuum dépassant trois ou quatre dimensions. Mais il n'est pas interdit de rêver qu'on puisse un jour transférer sur cartes perforées toute la documentation disponible au sujet des sociétés ..., et démontrer à l'aide d'un ordinateur que l'ensemble de leurs structures techno-économiques, sociales, et religieuses ressemble à un vaste groupe de transformations.

and ß-logics. Is it not possible, taking account of all that has now been said here, that it is *within this context* that the numbers may most effectively 'support and arm us in our struggle with the kingdom of nothingness'? Is it not in this manner, the manner of technology in its most essential sense, that already 'the numbers have begun to function in another way'? Clearly we have to move beyond symbolic paradigms in our logics, and into the world of the subsymbolic paradigm.

Of course none of us here can be precise about where we are travelling and about how we are going to get there, but then that is also of the essence of that thinking activity which is called *research*.

3.3   Symbolic and subsymbolic paradigms

A very large part of AI is currently taken up with *symbolic processing*. Thus, for example, in Knowledge-Based Systems we have facts and rules, usually written in natural language, and so using the alphabet and other symbols, and we manipulate the corresponding verbs, adverbs, nouns, adjectives and so on in such a way that we can draw conclusions from these facts and rules, which conclusions come to appear in the same natural language. Indeed McCarthy (1979) has characterised the main features of the principal language in this field, of LISP, as follows:

1.   Computing with symbolic expressions rather than numbers; that is bit patterns in a computer memory and registers can stand for arbitrary symbols, not just those of arithmetic.
2.   List processsing, that is, representing data as linked-list structures in the machine and as multilevel lists on paper.
3.   Control structure based on the composition of functions to form more complex functions.
4.   Recursion as a way to describe processes and problems.
5.   Representation of LISP programmes internally as linked lists and externally as multilevel lists, that is, in the same form as all data are represented.
6.   The function EVAL, written in LISP itself, serves as an interpreter for LISP and as a formal definition of the language.

These characteristics of LISP serve to characterise symbol manipulation codes generally: there occurs, in effect, a 'play of language' which, although it can be very impressive and quite convincing, is never really anything else than this. It is this approach that we call generally the *symbolic paradigm*.

Now it is a common reaction of those who realise that, so to speak, 'there is nothing behind the words', or that 'it is all a play on words' — that the code has no 'understanding' at all of 'what it is saying' — that they become rather contemptuous of this approach. The whole thing takes on the appearance of a confidence trick (see Fishman, 1981, pp. 347-349). For example, it is often said that one could better speak of 'pseudo intelligence' than 'artificial intelligence'. However, this cannot be a denigration of the symbolic paradigm without being a denigration of the standard mode of human discourse. Indeed, in a celebrated passage, Heidegger (1927/1962, pp. 212, 213) described this standard mode in just this way as follows:

> In the language which is spoken when one expresses oneself their lies an average intelligibility; and in accordance with this intelligibility the discourse which is communicated can be understood to a considerable extent, even if the hearer does not bring himself into such a kind of Being towards what the discourse is about as to have a primordial understanding of it. We do not so much understand the entities which are talked about; we already are listening only to what is said-in-the-talk as such — what is said-in-the-talk gets understood; but what the talk is about is understood only approximately and superficially.
>
> ... What is said-in-the-talk as such, spreads in wider circles and takes on an authoritative character. Things are so because one says so. The average understanding ... will *never be able* to decide what has been drawn from primordial sources with a struggle and how much is just gossip. The average understanding, moreover, will not want any such distinction and does not need it, because, of course, it understands everything.

It will be clear, however, that this 'what is said-in-the-talk as such' is precisely that which is expressed in the common-sense truths — in the rules, contracts, legislation and so on — which enter into hydroinformatics systems. Accordingly, the symbolic paradigm is an essential part of hydroinformatics. It is no more, even if no less, of a 'confidence trick' than are the artefacts which it represents.

This paradigm clearly still has considerable limitations, however. For example, in most existing KBS technology, the interrogation of the user can proceed only within a set context (such as that wherein the user is asked to respond to questions, commonly by moving a cursor to a particular answer among a list of screen-displayed, possible answers: see Hayes-Roth *et al*, 1983) and the responses of the user can only be used by the system within this same context. Thus, in most of the existing KBS technology, the difference between monotonic reasoning (in which new knowledge does not intervene during the reasoning process) and non-monotonic reasoning (in which it does intervene) is nothing like so great as it could be in principle. There is no way, for example, that the usual KBS can accept, as a response to a query, the answer stating that the correct reply is something quite other than anything listed or 'comprehended' by the KBS itself. In effect, the KBS cannot explain its own failures. Thus most existing KBS technology cannot even exploit all the possibilities inherent in explicit logical reasoning, lacking, as it does, any 'deeper knowledge', let alone 'primordial understanding', of 'the entities which are talked about'. Of course, one can try to generalise the knowledge base by providing some kind of 'deep knowledge' explicitly (e.g. Engelmore and Morgan, 1988, and, separately, Hayes-Roth, 1985), but this can provide, at best, only a limited (even if still useful) increase in range of applicability. It will then be clear that the symbolic-paradigm kind of non-monotocity is, at best, barely adequate for ecological studies, while it must become inadequate as soon as hydroinformatics interacts with social paradigms.

The response of the AI community itself to these limitation of KBSs has been described by Smolensky (Rumelhart *et al* 1987/1988, p. 197) as follows:

> The vast majority of cognitive processing lies between the highest cognitive levels of explicit logical reasoning and the lowest levels of sensory processing. Descriptions of processing at the extremes are relatively well-informed — on the high end by formal logic and on the low end by natural science. In the middle lies a conceptual abyss. How are we to conceptualize cognitive processing in this abyss?

> The strategy of the symbolic paradigm is to conceptualize processing in the intermediate levels as symbol manipulation. Other kinds of processing are viewed as limited to extremely low levels of sensory and motor processing. Thus symbolic theorists climb *down* into the abyss, clutching a rope of symbolic logic anchored at the top, hoping it will stretch all the way to the bottom of the abyss.

> The subsymbolic paradigm takes the opposite view, that intermediate processing mechanisms are of the same kind as perceptual processing mechanisms. Logic and symbol manipulation are viewed as appropriate descriptions only of the few cognitive processes that explicitly involve logical reasoning. Subsymbolic theorists climb *up* into the abyss on a perceptual

ladder anchored at the bottom, hoping it will extend all the way to the top of the abyss.

Now 'ß-logics' have to do primarily with *completely-connected systems* of thought, whereby new concepts can only be introduced so long as connections are already in place to accept them and so long as they do not contradict, or otherwise are consistent with, existing concepts and relations. In 'α-logics', on the other hand, we commonly observe only disparate 'insights', 'hunches', 'gut-feelings', and such-like fragmentary impressions, most of which cannot even be formulated precisely as concepts. We are for the greater part in a world of broken thoughts and utterances.[15] When describing the relevance of the computer productions of the subsymbolic paradigm to this, α-kind of thinking, Smolensky (Rumelhart *et al*, 1988, p. 252) wrote:

> The point is that such *productions are just descriptive entities; they are not stored, precompiled and fed through a formal inference engine*: rather they are *dynamically created* at the time they are needed by the appropriate collective action of the small knowledge atoms. Old patterns that have been stored through experience can be recombined in completely novel ways, giving the appearance that productions had been precompiled even though a particular condition/action pair had never before been performed.

> ... And since the productions are created on-line by combining many small pieces of stored knowledge, the set of available productions has a size that is an exponential function of the number of knowledge atoms. The exponential explosion of compiled productions is virtual, not precompiled and stored.

When faced in 1980 with the problem of how AI could best be applied within the nuclear industry, Abbott *et al* (1983) concluded that it could best concentrate on the learning function in game-playing-like situations. This approach effectively excluded symbolic manipulation techniques, so that

---

[15]There is a tendency on the part of ß-people to denigrate α-logic and even to express contempt for it. This antagonism appears, however, to be based more on an observed *misuse* of α-logic, such as is indeed so much easier to realise than in the case of ß-logic. Whatever the relative merits or demerits of such logics may be, however, it may make ß-people more balanced in their estimation (or increase their outrage further!) to recall that, on average, societies in all ages have provided much higher material rewards for the exercise of α-logics than they have done for the exercise of ß-logics. It is thus no accident which of these is associated with the letter α, and which comes naturally after it, with the letter ß.

attention had to be directed to non-symbolic, and indeed subsymbolic, approaches. Thus, although we had read Winston's (1977) pioneering work, we did not so much as mention symbol manipulation. However, outside of Japan (*e.g.* Gofuku *et al*, 1986; Wakabayashi *et al*, 1985), this approach was not replicated. The principal European nuclear consultants, strongly under the influence of academe, moved almost exclusively into symbolic manipulation codes, and this movement was further accompanied and strengthened by a number of ESPRIT projects which advanced in this same direction (KRITIC, BSB, DAIDA, RUBRIC: see CEC, 1988).

However, the ecological interaction process is the game-playing-like situation *par excellence*, and, of course, game-playing is then the learning process *par excellence* also (*e.g.* Margalef, 1968). Thus, although symbolic paradigms are essential in many hydroinformatic applications, and even in many ecological applications, if only as means of organising our own and our client's work and understanding, the use of subsymbolic paradigms is ultimately unavoidable.

At the same time, when seen within this context, the current principal subsymbolic paradigm of neural networking must itself appear exceedingly simplistic and mechanically materialistic when compared with the models of biological intelligence that are provided by the quantum mechanics of electro-chemistry and its field theory. Thus, field-like interactions are ignored completely in current neural networks, and 'frequency' effects, such as those which provide three-dimensional effects, cannot be accommodated at all. Moreover, from the point of view of 'depth' psychology, there is a serious mismatch between an electronic device in which very similar patterns of connection represent very closely-related patterns in our outer world of appearances, and living beings in which some of the most closely-related patterns appear to represent outer-world entities that are so far removed in semantic context, the one from the other, that the likelihood of their simultaneous appearance in our field of sensation is vanishingly small. The one approach brings neural syntax and semantics as close as possible together; the other often appears to set them as far as possible apart.[16]

---

[16]In my *magnum opus* I have attempted to combine the neural network metaphor with a 'minimum-number-of-neurons' condition to make plausible, in a modern-scientific way, the conceptual correspondences employed so consistently in alchemy and astrology. This same analogy then leads to a modern-scientific interpretation of the alchemical maxim, *obscurum per obscurius, ignotum per ignotius* (from the obscure to the more obscure, from the unknown to the more unknown). In the same vein, it must be obvious from every point of view (philosophical, anthropological, mathematical-logical, etc.), that no digital machine can possibly emulate the human mind in all of its functionalities: in this case Penrose (1989) has provided a whole book by way of demonstration — albeit one written almost exclusively from the ß-scientific point of view of quantum physics and related mathematics.

Is there not, as always, a deeper significance in all this? What, after all, are the connectionists really expressing in their *drive towards* subsymbolic paradigms, and in their corresponding *reaction against* symbolic paradigms? Is it not essentially this: that in the symbolic paradigm we 'block' the 'numbers' (0101101001, etc.) such that we can never obtain anything back from our system other than what we consciously introduced into it? Is it not that we have blocked-out the relation between our numbers and *their* outer world? Is it not that by doing this we prevent the numbers from telling us things of which we were not aware, even though we may ourselves have been at least the nominal authors of the models through which the numbers come to utterance in our modern world? Is not the connectionist movement a certain kind of *attempt to restore the autonomy of the numbers* — such as appears only to be possible within a subsymbolic paradigm? (See also, in this respect, the book edited by Eckmiller, 1990 and especially the introduction of Bachen *et al*, pp. 102-119).

If we see the matter in this way, then we see again how 'the numbers are beginning to function in another way'. Through the subsymbolic paradigm it appears that learning processes can be accommodated within what can always be construed as sets of numbers, so that the numbers attain an additional independence from our preconceived ideas — a certain 'autonomy', and so they attain to additional means to correct and instruct our own thinking.[17]

When contrasting the symbolic and subsymbolic paradigms from the point of view of neural networks, Smolensky explained the advantages of the subsymbolic approach in much the same way, as follows (Rumelhart *et al*, 1987/1988, p. 261):

Nowhere is the contrast between the symbolic and subsymbolic approaches to cognition more dramatic than in learning. Learning a new concept in the symbolic approach entails

---

[17]In translations into English of Heidegger's German, the independent, autonomous, essential being of an entity, *das Sein*, is commonly rendered as 'Being', while its mere existence in our field of apprehension, *das Seiendes*, is rendered as 'entity' or 'being' (see, for example, Steiner, 1978). In the present case we need to speak of 'the Being of the numbers', which translates back to much better effect as *das Sein der Zahlen*. (In alchemical and astrological number symbolism, on the other hand, each individual number has its own Being.) Examples of experiences of Being used by Heidegger which are particularly apposite in the centennial year of 1990 in which this is written, are our experiences of Van Gogh's paintings of a chair, or a pair of shoes. These artefacts 'take on a life of their own' with a tremendous, mind-searing impact in Van Gogh's renditions. Indeed, if we follow Barth's (1960, p. 298) appreciation of Mozart, that 'he knew something about creation in its total goodness that neither the fathers of the Church nor our Reformers, neither the orthodox nor Liberals, ... and certainly not the Existentialists ... either know or can express and maintain as he did', then we must similarly say of Van Gogh that 'he knew something about Being that none of the "philosophers of Being" either know or can express and maintain as he did.'

creating something like a new schema. Because schemata are such large and complex knowledge structures, developing automatic procedures for generating them in original and flexible ways is extremely difficult.

In the subsymbolic account, by contrast, a new schema comes into being gradually, as the strengths of atoms slowly shift in response to environmental observation, and new groups of coherent atoms slowly gain important influence in the processing. During learning, there need never be any decision that 'now is the time to create and store a new schema.' Or rather, if such a decision is made, it is by the modeler *observing* the evolving cognitive system and not by the system itself.

Similarly there is never a time when the cognitive system decides 'now is the time to assign this meaning to this endogenous feature.' Rather, the strengths of all the atoms that connect to the given endogenous feature slowly shift, and with it the 'meaning' of the feature. Eventually, the atoms that emerge with dominant strength may create a network ..., and the modeler observing the system may say 'this feature means the letter *A* and this feature the word *ABLE*'. Then again, some completely different representation may emerge.

The reason that learning procedures can be derived for subsymbolic systems, and their properties mathematically analysed, is that in these systems knowledge representations are extremely impoverished. It is for this same reason that they are so hard for us to program. It is therefore in the domain of learning, more than any other, that the potential seems greatest for the subsymbolic paradigm to offer new insights into cognition.

3.4   Logics and numerics: the third cycle in the characterisation of hydroinformatics

We may thus see in the subsymbolic paradigm of current AI an attempt to come to terms with what we have called α-logics by, so to say, 'freeing the numbers from the constraints of the large-grained symbols'. We are then in effect appealing to a certain sort of *autonomy of the numbers*, which I have called, within the context of CFD (Abbott and Basco, 1989), the *number myth*. In effect, then, we have to say that in the symbolic paradigm 'the large-grained symbols come to block the message of the myth'. However, when we place the matter in this light, we are in fact only repeating one of the oldest and most well-established of all dogmatic-scientific truths. This is the truism that the very *writing down* of a myth (whether legend, saga, or whatever other kind) is to reduce its mythological power. Indeed, in Greek, *mythos* means 'by word of mouth', so that a communication is strictly speaking no longer a myth as soon as it is written down. The word, once written, is reduced to mere *actuality*, while the spoken word remains open, in its very sounding, to any number of *possibilities*. We could then just as well say of the thoughts behind the words, that it is only when enunciated that *ils se relèvent; ils se réveillent à la vie.*

In the Western tradition this understanding subsists most clearly in Judaism. An excellent illustration is provided by Sutram (1972, pp. 63-66), as follows:

> La Tora que le juif a obtenue est une. Par sa nature, par ses origines, elle est orale. Il l'étudie et la transmet en l'accompagnant d'une mélodie. Les lois même dont la Tora abonde, qui constituent sa substance, ne se durcissent pas, car il chante; il les transpose dans le langage de la vie ...

> La Tora est, affirmons-nous, entièrement orale. Mais l'est-elle réellement? Elle se présente à nous dans sa forme écrite: non seulement dans sa partie dite 'écrite', dans la Bible, mais encore dans le Talmud ...

> Cependant, l'oral l'emporte sur l'écrit. Les textes bibliques se réfèrent ... à des 'paroles adressées, par Dieu', à l'homme et transmises atténuées, 'humanisées', par l'homme à son semblable. A l'origine du texte écrit est la parole divine; à son aboutissement, la parole humaine ...

> La Tora est incommensurable, insaisissable dans sa totalité. Les parchemins ne suffiraient pas pour la contenir. A l'instar de l'âme humaine, immatérielle, impalpable, elle ne saurait être localisée. Son siège est la mémoire: elle la garde et la féconde, autorise ses mouvements audacieux et soudains. La mémoire ne se réduit pas au fonctionnement, aussi complexe soit-il, d'un mécanisme cérébral, permettant de conserver une certaine matière qui s'y trouve déposée. La mémoire est aussi conscience: elle est, comme se la représente Bergson,

dynamique. Elle détermine l'utilisation judicieuse, efficace, de la matière qu'elle a enregistrée.

Les résultats des opérations de la mémoire pourraient aisément être mis par écrit: il serviraient mieux ainsi à leur application pratique. Pourtant la Tradition s'y oppose fermement. Car elle veut éviter l'immobilité, la pétrification.[18]

Even more explicitly, in the introduction to his translation of the *Dhammapāda*, Hanschandra Kaviratna explained the long-standing nature of this aversion to written text in all traditions as follows:

The art of writing, therefore, did not become popular, as the emphasis of education was on the development of memory and retentive power ... Paleographic evidence indicates also that writing, in its earlier stages, was mostly used to chronicle historic events, it was not used to impart instruction in mysticism and philosophy, exorcism and religion, for Druid bard and Brahman sage alike considered this a profanation of esoteric wisdom. In that golden epoch of intuition and memory culture no teacher ever attempted to hand down the sacred knowledge through the medium of script.

---

[18]*Author's translation:* The Tora that the Jew has obtained is a unity. By its nature, by its origins, it is spoken. He studies it and transmits it to the accompaniment of a melody. Even the laws, with which the Tora abounds, which are its substance, are not expressed harshly, for they sing; he transposes them into the language of life ...

We state that the Tora is entirely oral. But is it really? It is presented to us in a written form, not only in the part that is called 'written', in the Bible, but also in the Talmud ...

However, the oral preponderates over the written. The Biblical texts refer ... to 'words addressed by God' to man and transmitted as attenuated and 'humanised' by man speaking to man. At its origin the written text has the divine word; at its conclusion the word of man.

The Tora is incommensurable, not to be grasped in its totality. The parchments would not suffice to contain it. Just like the soul of man, it is immaterial and impalpable, it is impossible to localise it. Its place is in the memory which maintains it and fertilises it, authorising its sudden audacious movements. The workings of the memory cannot be reduced to the functioning of a mechanism — no matter how complex this may be, permitting the simple conservation of a certain amount of material deposited in it. The memory is itself conscious: it is, as Bergson represented it, dynamic. It determines itself the most judicious and effective use of the material that is consigned to it.

The results of the memory operations themselves could be easily put down in writing; this would even serve better for their practical applications. However, the tradition firmly opposes this. For it wants to avoid immobility, petrification.

In this view, the mere reading of the symbolic expression gives rise to a much greater risk of registering only 'what-is-said-in-the-talk as such', whereby what the talk is really about comes to be understood only indistinctly and superficially, if at all. When the message is maintained in the cells of the brain, in, so to say, the individual brain's own 'neural network', so it must necessarily remain so much closer to consciousness; while, when it is recalled directly from that place and transformed into sound, it is much more likely also to be sounded authentically. By these means it can be said again, but now of the thought contents generally, that, when sounded, *ils se relèvent, ils se réveillent à la vie*.

In a similar vein, in a standard *University Grammar of English* (Quirk and Greenbaum, 1973), it is explained that 'many of the devices we use to transmit language by speech (stress, rhythm, intonation, tempo, for example) are impossible to represent with the crudely simple repertoire of conventional orthography'. For statements to a similar effect to the above by French authors, ranging from Condillac, through J-J. Rousseau, to Saussure, see Derrida (1972/1982, pp. 139-153).

Thus, although the introduction of symbolic language obviously constituted a great step forward in the recording of human knowledge, this was necessarily purchased at a price, and this price was at its highest in the areas of legend and saga and other expressions of the deepest psychic experience, and thus in the areas of allegorical expression generally. All that is now being said is that, entirely analogously, the symbolic paradigm of AI, despite its great utility in more mundane but practically important areas of application, constitutes a great hindrance as soon as we wish to give expression to the deeper potentialities of, in our case, the number myth, such as is necessary for ecological and human-social applications. Speaking in metaphors — as we must at this level of discourse — we may describe the written text as 'frozen allegory' and the 'large-grained' symbols of symbolic AI as constituting 'frozen blocks' that quite prevent the flow of understanding that we most need.

Speaking from the side of quantum physics, Finkelstein (1978, p. 99) expressed this matter as follows:

> We choose as symbols physical systems that behave predictably, and so symbols come close to having the aspect of eternity. Logos supposes a perfect match between symbolic processes and physical, like that once supposed between the laws of Newton and the planetary process. The logos is eternal, the process transient, but somehow they match.
>
> In mythos, which is older than logos, process, including our involvement in it, is primary, and symbols are a kind of

demonstrative accompaniment, like chants accompanying ancient rites, declaring the rhythm and the values of the process but never defining it.

The historic transition from mythos to logos seems part of a pre-Platonic passage from the oral to the written tradition. Today the tide reverses.

If 17th century science evolves out of 16th century magic to minimize the mythical, romantic, illogical, and arbitrary in the world, then the greatest surprise of science must be that the actual world is to an overwhelming extent illogical, irrational and arbitrary. The illusion of rationality that fostered classical physics turns out to be the law of large numbers, which is not even a law in the sense of logos. Twentieth century psycho-analysis and quantum physics rediscover the enormous unperceived irrational content of the mental and physical process.

As I tried to explain in the third digression to the *Computational Fluid Dynamics*, the number myth that in earlier scientific eras took the outer form of an individual-number symbolism of alchemy and astrology today takes the outer form of a set-of-numbers symbolism, in numerical modelling. In both cases we have to do with a certain autonomy of the numbers, whereby they proceed on certain courses independently of our will. As observed earlier here, the most dramatic example of this process in computational hydraulics, and in computational fluid dynamics generally, is that of numerical instability, and indeed this case is so significant that I have gone back over it in some detail in the appendix that concludes this chapter. In CFD we observe generally how the model teaches us things of which we were not aware in our own minds. We descry 'a logic behind the numbers': the arrow in our Figure 17 that relates the model directly to 'its' outer world. And then, as is explained again and emphasised in the appendix to this chapter, *we cannot understand this relation scientifically*. That the model 'understands' its outer world in the way that it does is *something that we can always describe scientifically but never thereby truly understand*.

Now we are saying much the same thing in relation to the subsymbolic paradigm of AI: that it provides the means — or at least the promise of the means — to effect an α-logic on a digital machine. The very fact that we can effect numerical instabilities while working in a first-order language shows that the numbers can 'speak to us' in this language; and this fact alone gives the promise, if not the assurance, that the α-logic may also be realised as well in a first-order language. The question only remains: how can we best do this?

The number myth, being truly a myth, can only be brought to expression allegorically, such as in metaphors and parables, and numerical instability is one of the most potent of all such current metaphors. Now, however, we also seek allegories that are more generally embedded in logic, and this leads us to the subsymbolic paradigm. Having started by interpreting our strings of binary elements or 'bit patterns' (such as are represented by zeros and ones) as 'numbers', we persist in doing this also in the case of the elements of the subsymbolic-logic sets themselves, and we thus subsume all that we are trying to do here under the one rubric of the action of the number myth. Thus, our central object must appear as one of a raising up or a resurrection as a reification of the number myth, but now in its proper relation to the apparatus of modern science. Once yet again, we must say of the numbers that *ils se relèvent, ils réveillent à la vie*, so that they 'are beginning to function in another way'. In the words of Karl Jaspers (Jaspers and Bultmann, 1954, p. 20): *'Nicht Vernichtung, sondern Wiederherstellung der mythischen Sprache ist der Sinn.'* (*Author's translation*: the object is not the elimination, but the restoration of mythical language). But what then becomes of hydroinformatics? We see, correspondingly, that: *hydroinformatics is the coming-to-presence of the number myth as this now supports us in our stewardship of the arteries and veins of the biosphere.*

Of course it is pertinent to ask why we should need the whole massive edifice of hydroinformatics to achieve this. Since we have to do with a myth, we can only reply by referring to another myth. We then recall how in the concluding allegorical battle of the *Rāmāyaṇa*, set in what is now Sri Lanka, the two heroes are paralysed by the darts of an opposing Lankan magician, and how the monkey god, Hanuman, is sent to fetch herbs that will heal them from a hill far away in India. He travels there in next to no time at all, but then, lacking time to find the precise herbs to which allusion has been made, he brings back the entire hill; so that, even as he approaches Lanka, our heroes are released from their spell, and can return to the battle. Later a somewhat similar situation occurs, but now the herbs are on the top of a mountain. And now, for lack of time, Hanuman brings back the entire mountain! (The people of India are surely blessed that Hanuman did not have to return a third time!) The essential point is, then, that we need this vast apparatus of hydroinformatics because *we do not have time* to search for solutions in any other way.

It is then however pertinent to recall further that in concluding his translation of and commentaries on this same *Rāmāyaṇa*, Rajagopalachari (1952) posed the question of what *that* myth was essentially about, and he in his turn recalled the insight that it was about 'the relation between love and *Dharma*'. The best that we can do for this in English is to speak of a relation between love and 'predestination', accepting that 'predestination' has acquired such a mechanically-materialistic interpretation in the West that it has become little more than a caricature of *Dharma*. None the less, this is the best that we can at present do in setting the overall scene in which we recall the living presence of the number myth — other than that we must inevitably set this scene, composed by 'love and predestination', under the stronger and more sharply focused lights and other optics of modern natural science, as conditioned by our Judeo-Christian tradition.

This act of resurrecting the number myth as a living presence can only be achieved through love, while it would be to deny our *Dharma* if we refused to do this. We learn once again that the myth is indestructible: the numbers will not be denied!

The teleological question concerning numerical modelling has been posed in terms of how the numbers can strengthen the covenant between the Creator and His creature; but it will now be clear that this strengthening of such a 'mythically-expressed inexpressible' as the covenant can only be realised by another myth. It is only through the number myth that the relation between man and computer can transform from one of symbiosis to one of synergy. *Thus, at this level, we define hydroinformatics as the means whereby the number myth strengthens the covenant between the Creator and His creature in the realm of the arteries and the veins of the biosphere.*[19]

---

[19]Even Heidegger seems to have missed this impact of the digital machine. Indeed in his *Identity and Difference*, of 1969 (p. 41), he wrote:

> The time of thinking ... is different from the time of calculation that pulls our thinking in all directions. Today the computers calculate thousands of relations in one second. Despite their technical uses, they are inessential (*wesenlos*).

In fact, as must have been sensed already with the 1936 papers of Turing, on the one side, and Church, on the other side, a digital machine would necessarily introduce a whole new chapter in mythology. After all, the myth is just that which intercedes between our innermost existence, which is completely devoid of signs, and our outer world, which is, *qua* world, composed entirely of signs (of names and forms). That something should intercede between these two worlds — something that is of neither of these two worlds to the extent that it is a pure manipulator of signs, this alone was certain to have a profound influence upon current mythology. (See, more generally, Jaspers and Bultmann, 1954, and especially p. 77.)

3.5   Appendix: The logics of numerical instability and undecidability

3.5.1   Numerical instability and the number myth

As the notion that any myth might appear in this area at all must be quite alarming to many persons, and particularly so when this concerns a *number myth*, I shall try here to explain in some detail what I mean by this notion. But in the first place, a few words should be said about the individual numbers themselves. Surely, we might at first think, nothing could be so familiar as the numbers, nothing so obvious or well known as their rules of combination and permutation. And yet, as has been said on any number of occasions through the ages, and with particular emphasis in our own times, nothing is so mysterious as the numbers, and nothing so beyond our understanding as their ways of combination and permutation. We are constantly amazed, even spellbound, by the ways of the numbers, as indeed by the ways of mathematics generally. Indeed so often do we hear and read about the same wonder, the same sense of incredulity at the numerical result, that we scarcely know where to begin or end in quotation. For reasons of space we can take only a few examples. The first is then from the side of psychology (Jung, 1952/1955, pp. 57-58; for a discussion of the notion of 'the archetypes of the collective unconscious', see further Jung, 1943/1953, especially pp. 61-111):

> Since the remotest times men have used numbers to express meaningful coincidences, that is those that can be interpreted. There is something peculiar, we might even say mysterious about numbers. They have never been entirely robbed of their numinous aura. If, so a text-book of mathematics tells us, a group of objects is deprived of every single one of its properties or characteristics, there still remains, at the end, its *number*, which seems to indicate that number is something irreducible. (I am not concerned here with the logic of this mathematical argument, but only with its psychology!)

> ... Number helps more than anything else to bring order into the chaos of appearances. It is the predestined instrument for creating order, or for apprehending an already existing, but still unknown, regular arrangement or 'orderedness.' It may well be the most primitive element of order in the human mind, seeing that the numbers 1 to 4 occur with the greatest frequency and have the widest incidence. In other words, primitive patterns of order are mostly triads or tetrads. That numbers have an archetypal foundation is not, by the way, a conjecture of mine

but of certain mathematicians, as we shall see in due course. Hence it is not such an audacious conclusion after all if we define number psychologically as an *archetype of order* which has become conscious. Remarkably enough, the psychic pictures of wholeness which are spontaneously produced by the unconscious, the symbols of the self in mandala form, also have a mathematical structure. They are as a rule quaternities (or their multiples). These structures not only express order, they also create it. That is why they generally appear in times of psychic disorientation in order to compensate a chaotic state or as formulations of numinous experiences.

It must be emphasised yet again that they are not inventions of the conscious mind but are spontaneous products of the unconscious, as has been sufficiently shown by experience. Naturally the conscious mind can imitate these patterns of order, but such imitations do not prove that the originals are conscious inventions. From this it follows irrefutably that the unconscious uses number as an ordering factor.

Then, from the side of mathematics more generally (Huntley, 1970, pp. 142-143):

The unexpected meeting with the Fibonacci numbers in an improbable context such as a beehive, or in the expansion of an algebraic fraction ..., is another aspect of the pleasure of the discipline of mathematics. More than a feeling of pleased surprise at the sudden encounter with a familiar friend, there is a sense of amazement: 'How on earth did *you* get *here*?'

... Such emotions, archaic in origin, contribute to the charm of mathematics. The experience of glimpsing beauty in mathematics is as difficult to interpret to oneself as it is to communicate to a pupil. It is caught rather than taught. The student can only be encouraged to see the 'vision splendid' for himself. The joy, mediated through the intellect, originates in the lower strata of the mind, the arena of the emotions.

And then from the side of a philosophy that is quite highly critical of standard mathematical procedure (Wittgenstein, in Diamond, 1976, p. 112):

One asks such a thing as what mathematics is about — and someone replies that it is about numbers. Then someone comes along and says that it is not about numbers but about numerals; for numbers seem very mysterious things.

... Well, what is a number then? I can show you what a numeral is. But if I say it is a statement about numbers it seems as though we are introducing some new entity somewhere.

In fact this understanding has now penetrated so far that, for example, Cooke and Bez, in their *Computer Mathematics*, are led to exclaim (1984, p. 3):

The set A contained numbers and numbers are very strange — they do not exist!

The numbers cannot in fact be described in terms of other entities, but arise from out of our most immediate unverbalised experience, so they are correspondingly *prelinguistic*. Similarly, they cannot be derived logically from any other entities, whether of intuition or experience, so that they are prelogical or *prepredicative*. Since Husserl (1900, 1913/1970) it has been clear that such prelinguistic and prepredictive entities belong to the general category of the *prescientific*. Needless to say, this has not stopped any number of individuals from trying to 'define numbers' scientifically, but never satisfactorily. Wittgenstein (Diamond, 1976, p. 156) observed, already in 1939:

I want to get on to a terribly difficult business — a real morass — Russell's definition of number. It seems as though, if number is defined in this way (or Frege's way), everything will be clear. ... But ... this definition of number, which ought to make everything clear, won't bring us any further.

For his part, speaking from the side of mathematical logic, Manin (1977, p. 17) simply gave the Bourbakist 'definition' of the number 'one':

$$\tau_Z\Big((\exists u)(\exists U)\big(u = (U, \{\varnothing\}, Z) \wedge U \subset \{\varnothing\} \times Z \wedge (\forall x)((x \in \{\varnothing\})$$

$$\Rightarrow (\exists y)((x,y) \in U)) \wedge (\forall x)(\forall y)(\forall y')\big(((x,y \in U \wedge (x,y') \in U)$$

$$\Rightarrow (y = y')) \wedge (\forall y)((y \in Z) \Rightarrow (\exists x)((x,y) \in U)))\big)\Big).$$

and observed that 'it would take several tens of thousands of symbols to write out this term completely; this seems a little too much for "one"'. More definitively, we may recall Skolem's theorem, that 'no list of axioms in the symbolism of the first-order functional calculus can characterise the natural numbers categorically', while, 'if higher functional analysis is used, the deductive apparatus will be incomplete' (see Kleene, 1977, p. 638). We are here back essentially with the medieval point of view, that since the numbers belong to that which is common to both the Creator and His creature, the creature man, alone, can never understand them. Or, in the corresponding vocabulary of modern existentialism, number is an *existentiale*.

This must now suffice to indicate that our notion of number, at first sight so familiar, becomes ever more mysterious the more that we study it. This behaviour is, of course, inherent in everything that emerges from the level of the prescientific. It underlies, in particular, the number symbolism that provided the very bearing structures of alchemy and astrology (*e.g.* Jung, 1944/1953 and 1952/1955). In Abbott and Basco (1989), I have introduced this form of number symbolism as an earlier form of the number myth. Of course, one cannot expect any answer to a question like 'what is the number myth about?' — otherwise it would not need to be in the form of a myth in the first place. It can only, again, be 'caught rather than taught'. In the CFD book I tried to express it in the aphorism that 'the right way of numbers is the way of nature, and this absolutely, and so without exception'. This is, of course, scientifically meaningless; it is, indeed, worse than no definition at all, scientifically. But then we are no longer working 'at the level of the scientific'. We are here speaking essentially of Being, and every statement about Being must ultimately be couched in the language of the myth.[20]

The current form of the number myth arises from out of numerical modelling, so that it has to do with sets of numbers, and in practice with very

---

[20]Thence the 'robustness' of the myth relative to the language used to express it, just so long as this language is sounded authentically. Lévi-Strauss perceived this as follows (1958, p. 232):

Qu'on me permette d'ouvrir ici une brève parenthèse, pour illustrer, par une remarque, l'originalité qu'offre le mythe par rapport à tous les autres faits linguistiques. On pourrait définir le mythe comme ce mode de discours où la valeur de la formule *traduttore, traditore* tend pratiquement à zéro. A cet égard, la place du mythe, sur l'échelle des modes d'expression linguistique, est à l'opposé de la poésie, quoi qu'on ait pu dire pour les rapprocher. La poésie est une forme de langage extrêmement difficile à traduire dans une langue étrangère, et toute traduction entraîne de multiples déformations. Au contraire, la valeur du mythe comme mythe persiste, en dépit de la pire traduction. Quelle que soit notre ignorance de la langue et de la culture de la population où on l'a recueilli, un mythe est perçu comme mythe par tout lecteur, dans le monde entier.

large sets of numbers (commonly between 10^6 and 10^9 7-decimal-digit numbers). In view of its origins, we cannot be surprised if it takes any number of forms in our outer world of appearances. Its most dramatic form, however, is that of numerical instability, as illustrated here in Figure 18. We see how, within a period of a few prototype minutes, the water in a particular area (in this case the Venice lagoon) appears to have passed from possible velocities to entirely impossible velocities (of thousands of metres per second). We have here the numerical equivalent of a massive nuclear explosion. Now what can we say about this phenomenon of numerical instability scientifically, and how is it that this relates to the number myth?

We shall take the simplest possible examples, starting out from Chapter 1 of Abbott and Basco (1989). We there consider the random walk of a single 'molecule' between three contiguous cells, each of length Δx, over one interval of time, Δt. If the molecule is in any cell, let us say of address j, at time t, we then supposed that it can only move as far as cells with addresses $j - 1$ and $j+1$ during a subsequent time Δt. We denote the probability that it moves, on average, from address $(j - 1)$ to address j by $p\ (j - 1 \rightarrow j)$, the probability that it moves, on average, from address $(j+1)$ to address j by $p\ (j+1 \rightarrow j)$, so that the probability that, on average, it stays at cell address j is:

$$p(j \rightarrow j) = (1 - ((p(j-1 \rightarrow j) + p(j+1 \rightarrow j))$$

since there are no other possibilities. We now write as p_j^n the probability that the molecule is in cell address j at the time level $n\Delta t$. We then see that:

$$p_j^{n+1} = p_{j-1}^n\ p(j-1 \rightarrow j) + p_j^n\ ((1 - (p(j-1 \rightarrow j)$$
$$+ p(j+1 \rightarrow j)) + p_{j+1}^n\ p(j+1 \rightarrow j) \tag{1}$$

In words: the probability that the molecule is in cell address j at time level $(n+1)\Delta t$ is equal to the probability that it was in cell address $(j - 1)$ at time level $n\Delta t$ and moved one step to the right, plus the probability that it was in cell address j at time level $n\Delta t$ and stayed where it was, plus the probability that it was in cell address $(j+1)$ at time level $n\Delta t$ and moved one step to the left. The steps in this case are easily schematised, as shown in Abbott and Basco (1989, p. 4).

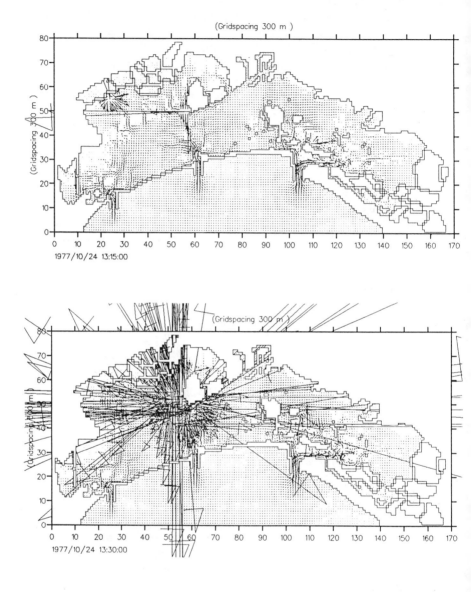

Note: Two stages in the 'explosion' of a numerical model due to numerical instability.

Figure 18   An illustration of numerical instability

Now if the fluid velocity corresponding to innumerable numbers of such motions is supposed to be a continuous function of x and t, it can be represented by a polynomial in x and t, which is to say that its variations can be described by a Taylor's series in x and t. Since, moreover, the fluid velocity is a consequence of molecular motions occurring with probabilities of the kind illustrated most simply in the above random walk, these probabilities can be supposed to exhibit the same variations as the fluid velocity which they provide. We can accordingly expand the probabilities in Taylor's series:

$$p_j^{n+1} = p_j^n + \left(\frac{\partial p}{\partial t}\right)_j^n \Delta t + \left(\frac{\partial^2 p}{\partial t^2}\right)_j^n \frac{\Delta t^2}{2} + (H.O.T.)$$

$$p_{j-1}^n = p_j^n - \left(\frac{\partial p}{\partial x}\right)_j^n \Delta x + \left(\frac{\partial^2 p}{\partial x^2}\right)_j^n \frac{\Delta x^2}{2} - (H.O.T.) \qquad (2)$$

$$p_{j+1}^n = p_j^n + \left(\frac{\partial p}{\partial x}\right)_j^n \Delta x + \left(\frac{\partial^2 p}{\partial x^2}\right)_j^n \frac{\Delta x^2}{2} + (H.O.T.)$$

where (H.O.T.) stands for 'higher order terms'. Substituting (2) into (1) then gives, at $j\Delta x$ and $n\Delta t$:

$$\frac{\partial p}{\partial t} + \frac{\partial^2 p}{\partial t^2} \frac{\Delta t}{2} + \frac{\Delta x}{\Delta t} p((j-1 \rightarrow j) - p(j+1 \rightarrow j)) \frac{\partial p}{\partial x} -$$
$$\frac{\Delta x^2}{2\Delta t} (p(j-1 \rightarrow j) + p(j+1 \rightarrow j)) \frac{\partial^2 p}{\partial x^2} = (H.O.T.) \qquad (3)$$

In the event that $p(j-1 \rightarrow j) = p(j+1 \rightarrow j) = \Delta p$, there is clearly no net motion and so no advection and in this case (3) reduces to:

$$\frac{\partial p}{\partial t} - \frac{\partial^2 p}{\partial t^2} \frac{\Delta t}{2} - \Delta p \frac{\Delta x^2}{\Delta t} \frac{\partial^2 p}{\partial x^2} = (H.O.T.)$$

In nature, $\Delta x/\Delta t$ must be of the order of the mean molecular velocity, which is finite, so that, strictly-logically speaking, $\Delta x^2/\Delta t \rightarrow 0$ as Δx, $\Delta t \rightarrow 0$. However, we can conceive of a sequence of 'equivalent walks' with $\Delta x^2/\Delta t$ kept constant and Δx, $\Delta t \rightarrow 0$, to obtain:

$$\frac{\partial p}{\partial t} + D \frac{\partial^2 p}{\partial x^2} = 0 \qquad (4)$$

where $D = \Delta p \Delta x^2 / \Delta t$. Since probabilities are non-negative real numbers, D is also a non-negative real number. This equation, then, describes a state of pure diffusion and (4) is correspondingly called the *diffusion equation*, with D the *diffusion coefficient*. Strictly-logically thinking, then, in this case the mean molecular velocity, $\Delta x / \Delta t$, → ∞ as Δx, Δt → 0^{21}.

In the event that all but one of the transfer probabilities is zero, such as $p\ (j \rightarrow j) = 0$ and $p\ (j+1 \rightarrow j) = 0$, so that $p\ (j - 1 \rightarrow j) = 1$ then there clearly occurs only an advection that coincides with the molecular transport. In this case (1) reduces to:

$$p_j^{n+1} = p_{j-1}^{n} \tag{5}$$

which expands, through term-by term cancellation of the H.O.T., to:

$$\frac{\partial p}{\partial t} + \frac{\Delta x}{\Delta t}\frac{\partial p}{\partial x} = 0$$

---

[21]Since $\Delta x / \Delta t$ corresponds to the molecular velocity, which is finite in any real-world process, when arriving at the limits, of $\Delta x \rightarrow 0$ and $\Delta t \rightarrow 0$, to obtain a partial differential form, we can only properly have $\partial p / \partial t = 0$, that is, a statement that 'nothing whatsoever happens'. What this means is that, strictly-logically thinking, one cannot describe a *diffusion* process with a *differential* equation: there is, quite simply, a disjunction in thought between the very concept of a diffusion process (averaging over a *finite* scale) and the very concept of a differential equation (a statement about what happens at a *point*). Accordingly, as pointed out in Abbott and Basco (1989, p. 371), to introduce a partial-differential equation to describe a diffusion process, although a standard practice in analysis for some two hundred years, is to perpetrate a logical absurdity. This appears as soon as we model such a process numerically. In particular, it is found that the diffusion coefficient, D, used in the model must itself be a function of the grid size used in the model, and indeed, since Smagorinski (1963; see also Leslie and Quarini, 1979) we conventionally take D proportional to Δx^2. Since then $D \rightarrow 0$, we find once again that there can be no diffusion effect at all engendered from the diffusion terms. This result, in its turn, opens up the further possibility that the numerical computations may come to reproduce the mixing processes observed in nature 'by themselves' or 'of their own accord', so to speak. The theory in this case, of large-eddy simulation (LES), was brought over in the first instance by Kraichnen, from quantum physics, and further elucidated, in particular, by Leslie (see again Leslie and Quarini, 1979). A sketch is given in Abbott and Basco (1989, pp. 360-364) while some of the practical consequences are detailed in the chapters of Bedford, Holly and Vieira in the reference book edited by myself (Abbott and Price, 1992).

There are also more numerical-analytical consequences, such as that whereby, strictly speaking, no stable explicit schemes can exist for parabolic equations such as (4), but such schemes can only function through the truncation errors of the schemes transforming them into equations of hyperbolic type. This in turn opens up the possibility for building explicit schemes that are unconditionally stable but then necessarily only conditionally consistent — the so-called 'schizophrenic' schemes (see, for example, Abbott and Basco, 1989, pp. 80 and 113).

If now $\Delta x/\Delta t$ is taken as constant as Δx and Δt tend to zero, so that $\Delta x/\Delta t$ represents the locally constant velocity of all the molecules, then we have:

$$\frac{\partial p}{\partial t} + u \, \frac{\partial p}{\partial x} = 0 \tag{6}$$

as the equation describing pure advection. In this case, u coincides with the (macroscopic) fluid velocity. Equation (6) is accordingly called the *advection equation* and also — and not incorrectly for this case where advection and transport coincide — the *transport equation*.

Thus, starting with a statement of finite length written in the first-order language of arithmetic, L_1Ar (even though the symbol p may stand for a real number), we arrive at 'equivalent' statements in the language of the continuum, which can be described in terms of the second-order language of real numbers, L_2Real.

Beginning with (4), we know that if we have initial conditions (at $t = 0$) of the type shown in Figure 19, the exact, analytic solution will develop in time as shown also in Figure 19.

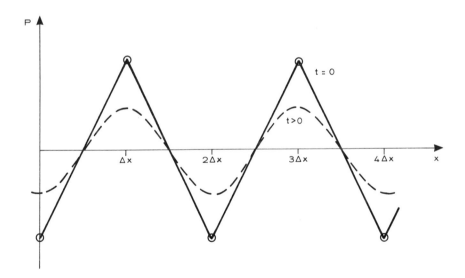

Figure 19   Analytic solution of a saw-tooth wave function

Now, however, we introduce $p\,(j-1 \to j) = p\,(j+1 \to j)$, $= \alpha$, into (1) so as to obtain (1) in the form of a numerical scheme:

$$p_j^{n+1} = \alpha p_{j-1}^n + (1-2\alpha)p_j^n + \alpha p_{j+1}^n \tag{7}$$

as our statement in $L_1 Ar$. Applying this to the initial conditions shown in Figure 19, in which:

$$p_{j-1}^o = p_{j+1}^o = -\,p_j^o \quad \text{for every} \quad j,$$

gives:

$$p_j^{n+1} = (1-4\alpha)p_j^n \tag{8}$$

Then, so long as $\alpha < \frac{1}{4}$, we have a behaviour that at least bears a resemblance to that shown in Figure 19, while for $\frac{1}{4} \leq \alpha \leq \frac{1}{2}$ we can still maintain:

$$|\,p_j^{n+1}\,| < |\,p_j^n\,|$$

even though the form alternates. However, for $\alpha \leq \frac{1}{2}$ we cannot even have:

$$|\,p_j^{n+1}\,| < |\,p_j^n\,|$$

Thus, as soon as $\alpha > \frac{1}{2}$, we have $|\,p_j^{n+1}\,| > |\,p_j^n\,|$, and thus instability.

But then we observe that $\alpha > \frac{1}{2}$ corresponds to $(1-2\alpha) < 0$, and thence, recalling that this $(1-2\alpha)$ corresponds in its turn to $p\,(j \to j)$, and so to a probability, we see that $\alpha > \frac{1}{2}$ corresponds to our introducing a 'negative probability', *i.e.* a *contradictio in adjecto*. This gives us a first intimation of the notion that, as introduced earlier, 'instability is the numbers' way of telling us that our scheme contains contradictory statements.'

However, all this *depends upon us knowing* that $(1-2\alpha)$ in (7) corresponds to a probability, so that it cannot then be a negative number. Because such interpretations are not usually available, we almost always proceed otherwise when *analysing* a scheme for stability. There are, again, any number of ways of proceeding in such an analysis. For example, we can picture a 'configuration space' which has the individual p_j^n as coordinates, so that an instantaneous state of the system (the sequence $\{p_j^n\}$) is mapped into a 'point' in the configuration space which moves under the influence of the numerical scheme. We can then ask under what conditions any two such points, say P and

Q, come closer together under the operation of the scheme, as schematised for the subspace composed by p_o, p_1 and p_2 in Figure 20.

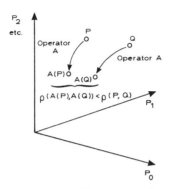

Figure 20   **Schematisation of a contraction mapping**

Thus, if we take the simplest case, defining the distance between two such points by $\rho(P,Q) = \text{Max} \mid p_j\text{-}q_j \mid$, where $\{p_j\}$ and $\{q_j\}$ are the coordinates of P and Q respectively, we find (Kolmogorov and Fomin, 1957, Vol. 1, p. 45) that in the case of a general finite linear operator, A, of the form:

$$q_i = \sum_{j-0}^{jj} a_{ij}\, p_j + b_1$$

we have the general contraction condition that:

$$\sum_{j-0}^{jj} \mid a_{ij} \mid\, < 1 \qquad\qquad (9)$$

In the case of (7) this gives:

$$\mid \alpha \mid + \mid 1\text{-}2\alpha \mid + \mid \alpha \mid\, < 1 \qquad\qquad (10)$$

as the required condition. Since stability implies only that such points do not get further apart (as follows from setting $\{p_j\} = \{p_j^n\}$ and $\{q_j\} = \{p_j^{n+1}\}$, then (10) can be relaxed to:

$$\mid \alpha \mid + \mid 1\text{-}2\alpha \mid + \mid \alpha \mid\, \leq 1 \qquad\qquad (11)$$

giving, again, $\alpha \leq \frac{1}{2}$. It is easily shown that this particular condition is necessary as well as sufficient. In this analysis, however, we have no inkling of a contradiction, and such *interpretations* of $(1-2\alpha)$ as a 'negative probability' *are entirely lost to view*. A similar experience is described in relation to mathematical proof theory generally by Lakatos (1976).

The same loss of interpretations occurs when we analyse (7) in the Von Neumann manner, in terms of the orthogonal basis functions $\{\exp 2\pi kj\Delta x/2l\}$ where k is the dimensionless wave-number index, and l the extent of the domain of the solution. We then find the relation between Fourier coefficients $p(k)^n$ and $p(k)^{n+1}$ (*e.g.* Abbott and Basco, 1989, p. 107):

$$p(k)^{n+1} = [1-4\alpha \ \sin^2 \ (\frac{2\pi k\Delta x}{2l})] \ p(k)^n \tag{12}$$

so that the Fourier coefficients can never increase in magnitude in time so long as $\alpha \leq \frac{1}{2}$. Once again, we gain a particular result, but lose a general interpretation: once again, there is not the slightest whiff of contradiction in the air.

Now it is a result of this loss of interpretation that *in practice hardly any use is made of stability analyses*. Of course these are derived and duly presented in publications, but that is a matter of scientific decorum and propriety and not of great practical utility. *In practice, in the real world of numerical modelling, the development of a modelling system proceeds experimentally, from one instability to the next, through a range of models*. In every case that we encounter an instability, we seek out the inner contradiction in our own reasoning that we have implanted in the code. We then find, *in every case*, that as soon as we have corrected this contradiction the particular instability no longer arises. The model that we are using can then run further — to the next instability, and so the next contradiction. In every case ever encountered to date, at least in my own experience and the experience of those with real-world modelling experience with whom I have spoken about this matter, an interpretation of instability has been found, in by far the greatest number of cases, in terms of such a 'logic behind the numbers'. Thus, in the real world of numerical modelling, most of our work is based upon this presupposition, which is essentially that of the number myth.

We thus constantly experience the situation in which (Abbott and Basco, 1989, p. 377) 'the model comes to teach us things of which we were previously unaware, even though the model is itself entirely a product of our own minds'. This is to say that the sets of numbers and the sets of arithmetic operations on these numbers that make up a model of some aspect of our physical world together appear to 'understand' the behaviour of this world in a more systematic manner than we do ourselves. The question that then naturally poses itself is whether we can find an interpretation of 'the way that the model does this'. In order to consider this question we can best pass on to our second example, of the advection equation. In order to see this we shall admit some numerical diffusion into the advection equation by letting $p \ (j - 1 \rightarrow j) = 2$, even while keeping $p \ (j+1 \rightarrow j) = 0$. We then have the scheme:

$$p_j^{n+1} = 2p_{j-1}^n - p_j^n \tag{13}$$

This is, of course, also consistent with $\partial p/\partial t + u\partial p/\partial x = 0$, but its truncation error contains a diffusion-like term.

As $p \ (j - 1 \rightarrow j) = 2$ implies that $p \ (j \rightarrow j) = -1$, <0, the scheme contains a further *contradictio in adjecto*, so that we can be sure that it will be unstable. In the same vein, its truncation error contains a 'negative diffusion coefficient', which provides a destabilising, antidiffusive behaviour. We then consider the initial conditions:

$$p_0^n = 0, \quad j = 1,2,... \ jj,$$

and the left boundary conditions:

$$p_0^n = 1, \quad n = 0,1,2,... \ nn.$$

The 'exact' solution (the solution of the equivalent statement in L_2Real, which in this case is given by (6)), is:

$$p_o^n = H(\frac{2\Delta x}{\Delta t} t - x)$$

with $H(x) = \begin{cases} 1, & x \geq 0 \\ 0, & x < 0 \end{cases}$

while the scheme (13) provides:

$$p_0^0 = 1, \quad p_1^1 = 2, \quad p_2^2 = 2^2, \quad ... \quad p_{2i}^{2i} = 2^i, \quad ... \quad p_{nn}^{nn} = 2^{nn}$$

We observe the usual feature of such an unstable scheme, that if we reduce the distance and time steps (such as by halving both of them) while keeping the same scheme (in this case (13)) the p_{nn}^{nn} increases for any fixed $T = nn\Delta t$ (in that, for example, $p_{nn}^{nn} \ (\Delta x, \Delta t) = 2^{nn}$, $p_{nn}^{nn} \ (\Delta x/2, \ \Delta t/2) = 2^{2nn}$).

We then have *Richtmyer's* (1957) *paradox*, which we can express as follows:

As we refine the mesh so that $\Delta x, \ \Delta t \rightarrow 0$, while keeping the same unstable scheme, the statement in L_1Ar (in this case (13))

satisfies ever more closely the solution of the equivalent statement in L_2Real (in this case, (6)) even while the solution of the
statement in L_1Ar departs further and further from the solution
of its equivalent statement in L_2Real.

This is then to say, in the present connection, that *the numbers insist on
being heard*: we can never escape hearing what they have to say simply by
refining our descriptive system. We can, of course, change the statement such
that the contradiction does not arise. For example, we can introduce a numerical
diffusion into a scheme such that an incipient instability is dissipated, but then
our scheme is describing what is in effect another situation, in which, due to
diffusion, the contradiction no longer arises. By raising their voices in this way,
'the numbers always tell the truth'.

The numerical-modelling-system builder accordingly tries as far as
possible to work with a modelling system that does not admit numerical
diffusion, or even numerical dispersion, so that any model that he uses to
develop the system can 'speak as clearly as possible'. After this, however, he
must appear as an incorrigible opportunist, trusting one formulation until, led
by instability, he develops another. In the words of Barth (1922/1931, p. 336):

> The prophet will adopt no particular point of view without the
> secret intention of abandoning it as soon as he has gained a
> merely tactical advantage; for there is no question but that his
> point of view will be shown to be finally inadequate.

To this we should add that this leading-on feature of instability takes on
another dimension again as soon as we combine numerical models with other
components of a hydroinformatics system. In particular, as soon as we introduce
an updating of model results on the basis of measured data, so that we come to
assimilate these data into the model, we may easily introduce contradictory
statements, and thus induce instability (Heemink, 1986; Heemink and
Kloosterhuis, 1990). A similar source of instability must be anticipated from the
introduction of logical constraint sets deriving from common-sense rules and
other heuristics. All such expectations and anticipations then arise from the
implicit assumption of the number myth.

We are here dealing with a myth, and since the time of the old Greeks,
at least, endless attempts have been made not just to 'explain the particular
myth' but to provide a methodology for doing this for all myths (see, again,
Dilthey, 1976). Such exercises belong to the art and the science of *exegesis*, and
the number myth also does not lack for exegetic investigations. The myth is an
allegory, and an allegory functions just so long as we do not come to experience
it as *only* an allegory. If *that* happens, then we are led to 'explain' it in other

terms, but since we are always trying 'to give expression to the inherently inexpressible' we always obtain 'only' another allegory. Exegesis then ends when we are no longer aware that what we now have is, itself, 'only' an allegory. This is to say that we can *explain* the number myth in any number of ways (and in principle in an infinity of ways), but we can never *thereby* understand it. We can only understand it at all 'on its own grounds' or 'for its own sake', or 'as such': only by making a 'leap of faith', not through rational argument (see, again, Jaspers and Bultmann, 1954). And then, of course; 'Faith is a miracle, otherwise it is not faith'.

Thus, in each and every case that instability arises, we can only wonder, like Huntley, 'how did *you* get *here*?' We can only wonder how these sets of numbers and operations 'come to find out' all the things that they do, so exhaustively and so completely. Of course *we* can always provide an explanation, after the event, such as the intervention of 'a negative probability' or, what is the same thing, a 'negative diffusion coefficient', but the more that we experience the manifestations of the number myth, the greater our mystification becomes. We are at one with the AI experience described earlier, from Kay (1984). We repeat:

> What does it mean to represent something from our world in a world that is not of our world? What does it mean to do things in that world that are not like the things in this world? How can we translate back into our world in such a fashion as to get a message that actually means something to us? I think that is mysterious to everyone. ... The reason for the mystery and fragility is not understood.

We should observe that the manifestations of the number myth are by no means limited to those of instability, and indeed it is an infallible guide to seek these manifestations wherever we find a paradox, such as the 'resolution paradox' of weak solutions of conservation laws or the 'circulation paradox' of rotational flows (Abbott and Basco, 1989). A further example that does not involve instability is illustrated in Figure 21.

In the same vein, the number myth will impress itself most strongly upon those who are most closely engaged in modelling the physical world: we should correspondingly not expect the number myth to attain to much credence among numerical analysts, who are usually further removed from contact with this outer world. Indeed, to the scientist, *per se*, the whole situation surrounding the number myth — this 'we do not understand!' — will appear as nothing less than a scandal, but then every attempt by man to intercede between the emanations of the deepest levels of his psyche and his outer world of name and form must appear in this light, as a scandal (Barth,1922/1931, pp. 332 *et seq*).[22] But then the number myth has almost nothing to do with science-as-

---

[22]This scandal has at its centre the scandal of absurdity, of paradox, and Christianity, as the creed of the absolute paradox of the God-man, the *Christos*, can never evade this absurdity: it is, preeminently, *credo quia absurdum* ('I believe it because it is absurd'). The paradigm was set already by Tertullian: 'And the Son of God is dead, which is worthy of belief because it is absurd. And when buried He rose again, which is certain because it is impossible.' As has now been described in any number of ways, and not least in the psychology of Jung, there is no way around this manner of expression of essential, eternal truth; and all the 'demythologising' in the world will not remove from it its sting of absurdity and its stigma of scandal. So long as we in our own minds keep to logical, 'linear' or 'one-dimensional' thinking, in the manner of the logical chaining of the inference engine of a knowledge-based system, we can never attain to any deeper truth. The 'compilers' and 'operating systems' of our own minds will then simply go on functioning within just this standard mode of operation and we can have no way of enquiring into their mode of functioning, let alone of changing this mode of functioning. It is only by addressing this, our 'systems software', in a manner for which it was not, so to say, 'designed' that we are able to penetrate into the layers that underlie its functioning. Accordingly, all the devices of magic, whereby the means of conscious thought and action are brought to bear upon the workings of the 'compilers' and 'operating systems' that are situated in the unconscious, must necessarily appear as illogical, absurd and paradoxical. Entirely consentaneously, any fault that is situated at the level of our unconscious 'operating system', as is the Biblical-Kierkegaardian 'sickness unto death', can only possibly be accessed and so treated in this manner. In Barth's work, this sickness has a causal agent, which receives the name of 'nothingness' (*das Nichtige*), following a tradition that goes back at least as far as Augustine. Then, to the extent that this agent makes its presence felt in our outer world, as evil, so it is posited to organise itself as a power, which then becomes 'the kingdom of nothingness'.

In similar but more lowly veins, every truly profound change or 'paradigm shift' in thought proceeds, to use Kierkegaard's expression, 'on the strength of absurdity' or 'by virtue of the absurd' (*i Kraft af det Absurde*). The paradigm in modern mathematics was Dedekind's *definition* of an infinite set as one that is equivalent to a proper subset of itself, and Cantor's subsequent introduction of transfinite induction, such as was thoroughly castigated for its absurdity — and rightly so — by almost all of Cantor's contemporaries. However, by 1977, Manin could write (p. 106): 'In the hundred years since the introduction of transfinite induction, not a single new method of constructing sets has come into common use'.

In the case of our usual programmable Von Neumann machine, on the other hand, we deliberately exclude the possibility of our code written, say, in Pascal, altering the Pascal compiler itself, let alone the possibility that this code could change the code of the operating system of the computing machine. In fact, our Von Neumann-type machines are *designed just to exclude this possibility*. Although this is of course essential from a practical computer-scientific point of view, it does on the other hand have the effect that we, so to speak, 'code ourselves into a corner' relative to our own, natural abilities. It is the dawning realisation of this situation that has led to the 'revolt' in some quarters against the Von Neumann machine, and so it is this which has promoted, correspondingly, such devices as neural computers. Of course, having coded ourselves into a corner at one level, we can still try to code ourselves out of it again at another, higher level, such as by building neural-network emulators on Von Neumann-type machines. There are, for example, two very substantial ESPRIT projects (EP2059 PYGMALION and EP2092 ANNIE) that are in part concerned with just this kind of approach (see again Eckmiller, 1990, but now pp. 477-483 and pp. 373-380). It is all monstrously inefficient in machine computing resources, of course, but, given the ever-falling processing costs of the basic

such, even as it has everything to do with technology. It is just that 'other way' in which 'the numbers are beginning to function in another way'.

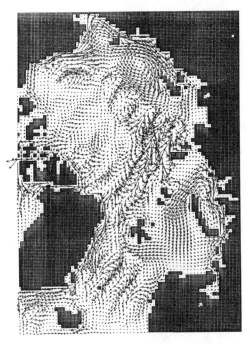

Note: The example shows flows in a three-dimensional model of the Arctic Ocean. In (a) a distinctive and highly unnatural 'curdling' of the flow is observed.

Figure 21   'The model tries to tell us something'

---

hardware, this probably does not matter so very much in practice. What is very interesting in this respect, however, is to observe such devices in operation over the often long periods that they need to settle down into stable patterns of behaviour. One is unavoidably impressed by their apparent *repeating of the same experiment*, time and time again, *ad infinitum et ad nauseam*. One is irresistibly reminded of the practice of alchemy, where again apparently the same experiment was performed over and over again — and thence of Wolfgang Pauli's observation, made in just this context, but of course from the side of quantum physics, that there is in reality no such thing as 'the same experiment.'

It should then only be added here that the further absurdity of the *scientific* study of the *credo quia absurdum* is by no means restricted to dogmatic science, but can be followed in many other places, as exemplified in the Abbott and Basco *Computational Fluid Dynamics* and, as an amplification of one of the aspects introduced there, of the Fourier equation, earlier in this section. To the extent that Jung's psychology follows this same approach — and in this respect it is fully compatible with Christian Church dogmatics — it is itself every bit as absurd and every bit as scandalous. And there is just no way around this.

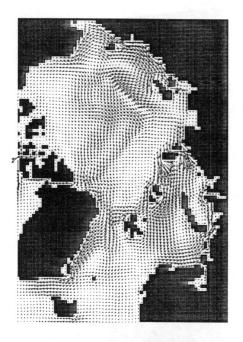

(b) Changing the fluid pseudo com-
pressibility cures the problem, showing
that it is a compressibility-dependent
effect (in this case interacting with the
Coriolis acceleration).

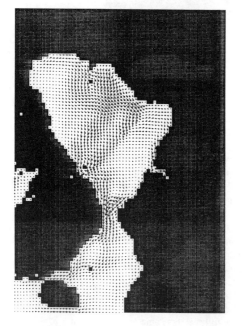

(c) But then a view of the flow at 500
m depth shows a 'stiffening' or
'freezing' of the flow along the
directions of the numerical grid,
showing that reducing the com-
pressibility is not enough, but that
correction terms must be introduced in
the model to account for the effect so
discovered.

Figure 21   (continued)

3.5.2   The limits of hydroinformatics: undecidability, non-computability and a return to the number myth; the influence on engineering education

We have seen that hydroinformatics brings together both deducing and computing aspects of digital machine utilisation, and indeed, from the narrow, scientific-content point of view of process modelling, hydroinformatics can be supposed to be composed by the union of artificial intelligence and computational hydraulics. It follows, by way of example, that its mathematics must have a bias towards the study of recursion, whether this takes the form of forward and backward chaining in the inference engine of an expert system shell or takes the form of forward, backward and double-sweep algorithmic structures in numerical modelling. As the general approach represented by this aspect of recursion is already well established in mathematical logic (*e.g.* Lightstone, 1978; Lyndon, 1964; Mendelson, 1964; Manin, 1977), while the way between this basis and AI has been set out by Genesereth and Nilsson (1988) and Thayse (1988) and the way between it and computational hydraulics by Abbott and Basco (1989), it is clear at once that mathematical logic provides, together with its consort of set theory, the fundamental mathematical basis of the science of hydroinformatics. Let us however now look a little further at these mathematical foundations.

Modern science, as Heidegger in particular explained, is mathematical through and through, even when it is not cast explicitly in mathematical form. Correspondingly, few engineers are aware that they are using mathematics at every stage of their careers, even though they are not actually 'writing equations'. But then, given that mathematical education in engineering generally lags about one hundred years behind the research upon which it is based, it is quite easy to predict the mathematical-educational requirements of the hydraulic engineer, even of the year 2000, simply by referring back to the state of mathematical research in the year 1900. This makes the task of predicting the needs of engineering mathematics particularly simple, since at the International Congress of Mathematicians held in Paris in 1900, David Hilbert presented 23 problems which he believed would occupy mathematicians throughout the entire 20th century. A survey made in 1975 (Browder, 1976) demonstrated that Hilbert's estimate was correct in the case of mathematics as such, even as the perception of the nature of these problems appeared to have changed as the century progressed. However, over the period 1970-1990, some of these Hilbert problems have also come to influence technologies like hydraulics. It must suffice here to mention only the influences on hydraulics of the second of these Hilbert problems, but then in a little detail. Let us first state what this problem is.

Hilbert's second problem has to do nominally with the consistency of the axioms of arithmetic. Kreisel (1976) stated it in the form of a proposition:

It should be possible to exploit the finiteness of (all) proofs and
so establish the consistency of the axioms of arithmetic without
the use of (familiar) infinite models; even though the theorems
proved are ordinarily intended to be about infinite sets.

The attempt to verify this supposition, to demonstrate its general validity, came
to be called 'Hilbert's programme'. It was the intention to demonstrate that the
whole realm of mathematics could be constructed from out of a finite set of
formal axioms using only finite sets of formal-logical operations. In particular,
it should by these means be deducible, or *decidable*, whether any given theorem
was true or false. Correspondingly, in time, Hilbert's programme came to
devolve upon the problem of establishing this possibility, so that it became a
'problem of decidability', or *Entscheidungsproblem*. As in mathematical logic
every problem of deducibility is mirrored by a problem of computability (*e.g.*
Manin, 1977), the *Entscheidungsproblem* could also be posed in terms of
demonstrating the computability of any 'function', *i.e.* the existence of an
algorithm for solving any given problem.

As is now rather widely known, Hilbert's programme was shown to be
untenable, at least in its most general form, already by Gödel, in 1931. It
suffices to recall here that in the proof of the first of the two theorems
enunciated by Gödel at that time (see Van Heijenoort, 1967, pp. 592-617), it
was demonstrated that within any consistent system of logic or mathematics
generally that was rich enough to encompass arithmetic, statements could be
made that could be neither proved nor disproved. Such statements could only
be proved or disproved within a system that was richer in its axioms; but then
within this sytem again statements could be found that could be neither proved
nor disproved. Thus, as Heppenheimer (1990) has expressed the matter:
'Gödel's work demonstrated limits to the ability of mathematics to answer
questions in its own fields.'

Now from one point of view — which was for a long time the dominant
point of view in mathematical logic — Gödel's 1931 theorems concluded an
epoch, which Van Heijenoort characterised as one of 'the arithmetisation of
logic'. However, a central issue during this process of arithmetisation was that
of the possibility of the constructing and solving of logical polynomials which
had solutions exclusively in the field of integral numbers. These are the so-
called 'diophantine equations'. (This issue was posed independently as the tenth
problem of Hilbert's 1900 list.) It was in fact shown by Matyasevich in 1970
(see Chaitin, 1987, and Heppenheimer, 1990) that there is no general procedure
even for establishing the existence of solutions of such equations: any theory of
diophantine equations will entail formally undecidable examples.

Perhaps better known nowadays is the posing of this problem in terms
of an idealised computer known as a 'Turing machine'. We may in this case

recall Turing's question of whether one could decide whether, for any arbitrary program, this machine would run on indefinitely or come to a halt. Turing showed that this question is also undecidable: no general method can exist for determining whether any program whatsoever will halt.

These problems, of diophantine analysis and general algorithmic halting, came together in a series of papers by Chaitin (*e.g.* 1982, 1987). In summing up this work, Chaitin wrote (1987):

> In conclusion, we have seen that proving whether particular exponential diophantine equations have finitely or infinitely many solutions, is absolutely intractable. Such questions escape the power of mathematical reasoning. This is a region in which mathematical truth has no discernible structure or pattern and appears to be completely random. These questions are completely beyond the power of human reasoning. Mathematics cannot deal with them.

We might then observe, in passing, that this same conclusion appears to arise in the case of turbulent flows: the problem of determining the flow obtaining in the limit, as $\Delta x_i \rightarrow 0, i = 1,2,3$, and $\Delta t \rightarrow 0$ appears also to be an undecidable problem (*e.g.* Abbott and Basco, 1989, p. 366 and earlier in this appendix. Much of this reappears in the theory of fractals and in so-called 'chaos theory', but we shall not go at all into these aspects.)

It is here again, then, that we run out of the limit of scientific thinking even when we allow ourselves the possibility of using the most powerful of all possible digital machines. At the same time, however, we believe that numerical solutions to such problems do in fact exist, and it is at this point, as indicated by Finkelstein (1977), that we run again into the realm of the mythical. As pointed out earlier, in the quotation from Finkelstein, this situation has long been recognised in the area of quantum physics, and indeed Chaitin, Heppenheimer and other mathematicians working in this area have acknowledged this earlier recognition.

We have also seen, however, that the manifestations of the number myth cannot be restricted to the side of computability: we have also seen its manifestations within studies of the limitations of symbolic paradigms in AI, and particularly in the formulation and justification of subsymbolic paradigms, such as those currently realised using neural network machines and their emulators. Since, however, the questions of the potentials and limitations of the respective paradigms can be reduced to (or 'modelled' by) arithmetic questions, so that they also fall within the ambit of the number myth, we must expect the problem of undecidability to present itself to hydroinformatics on its subsymbolic-logical side as well.

As these considerations are already beginning to attract the interest of the hydraulician, they must also concern those engaged in education in hydraulics, and so in hydroinformatics. For although the hydraulic engineer may come to be relieved from many routine design and management tasks by advances in tools, languages and environments, this engineer will require a much deeper understanding of certain fundamental matters in order to make proper use of these facilities. In particular, a rather thorough grounding in mathematical logic and set theory will be required, probably with the emphasis upon the mathematical-logical side. These are, for example, foundational subjects in the Course on Hydroinformatics which is due to begin at the International Institute in Delft in October 1991. In view of what has just been said above, it should be explained that it is not intended in the Delft course to teach diophantine analysis, while the discussions of undecidability will be restricted to those arising from the Turing problem. The symbolic paradigm side will be advanced roughly along the lines already set out by Genesereth and Nilsson (1988). The mathematical-logical distinctions between (first- and second-order) languages will similarly be used to initiate the computational-fluid-dynamic side, as set out by Abbott and Basco (1989). The limitations of both approaches in ecological modelling, including its anthropological aspects, will then be associated with limitations arising from the more fundamental assumptions, leading to the use of the subsymbolic paradigm, as introduced in this work.

Over and above all these matters, however, the engineer of the future will need some guidance and advice on the subject of mythology. The nature of mythology needs to be explained: its inevitability in all thinking about society and science, and indeed in thought generally, its forms or metaphors, its connections, teleologies and doctrines, its hermeneutics and various exegeses: all of this needs to be introduced. The results of such as Dilthey, Husserl, Heidegger and Jaspers on the metaphysical side, and of such as Schleiermacher, Kierkegaard, Barth and Bultmann (with apologies for the Protestant bias!), on the other side, need to be communicated. We dare not leave our young engineers as helpless 'babes in the woods' when it comes to such critical and potentially dangerous matters as these. With such an understanding in place, on the other hand, the relation between man and his most fantastic creation, his digital machine, can transform from one of symbiosis to one of synergy. This is, of course, a necessary condition for facing the environmental challenges of the next century, and so for redressing the balance between the forces on the side of the creation and those opposing the creation, in the realms of the arteries and veins of the biosphere. It is high time to reduce the distance between the *Ecole des Ponts et Chaussées*, where these things are not currently taught, and the *Ecole Normale Supérieure*, where they presumably are (*e.g.* Derrida, 1972/1982).

Concluding unscientific postscript

Pondus meum amor meus, eo feror quocumque feror (My love is my weight; by it I am carried wherever I am carried).

Augustine, Confessions

I have drawn a thread through this book, which is that 'the numbers have begun to function in another way', and I have connected this to a number mythology that was supposed to have become extinct with the eradication of the number symbolism of the alchemists and the astrologers. As this eradication was the work of the church and modern science collectively, this appears to place me in the particularly invidious position of contributing to the resurrection of a myth which is incompatible both with religion and science. My first purpose in this postscript is to explain why I believe that there is in fact no such incompatibility. After that, I have to explain why I believe this approach to be necessary.

In its own way, my position here is by no means unknown to modern science itself. For example, it is well known how, in his study of the infinite, Cantor ransacked the houses of theology and scholastic philosophy for precepts and prototypes — and he certainly found plenty to occupy him (see, for example, Gilson, 1950). Similarly, Gödel's inestimable contributions to mathematics were inseparable from his very profound spiritual development, which came to merge upon the mystical. As has been repeatedly observed in this book, however, the number myth cannot be comprehended in terms of modern science, but is of an essentially technological nature. Indeed, not only is this number myth in no way scientific, but it is not even of the same nature as the myths of science, even though it may be contiguous to these. But where then is technology placed in respect to its myths? Having devoted my major (unpublished) work to such matters as this, I can only sketch the position of technology as this appears in hydroinformatics in the most partial and fragmentary way; but clearly this is necessary to the proper conclusion of this book, and so it must be attempted.

The position of this 'post-modern' technology can in fact be adduced from the logics that have already been introduced in order, so to say, to 'bracket' hydroinformatics. These were introduced as the ß-logic of modern natural science, which should need little introduction in this place, and an α-logic that, as it reaches its apogee in the dogmatic science of theology, is likely to be less familiar to most readers of this book. We have already seen that, in order to give expression to the full essence of hydroinformatics, we have been obliged to alternate between the vocabularies and thought world of natural science and the vocabulary and thought world of dogmatic science, and this already indicates that this technology is placed between that which is essential to the language and thought world of natural science and that which is essential to the language and thought world of dogmatic science. We may then regard dogmatic science, in turn, as the science of religious experience as this is worked out within the context of a particular social religion (see, for example, Schleiermacher, 1831/1976: in our present schemata, each social religion is then the means of shaping the religious consciousness to particular social ends). We might then say, further, that this technology occupies the space between what is essential to science and what is essential to that religious consciousness or awareness that is the dominant concern of social religion. When seen from this perspective, the technology with which we are here occupied is that which spans the space between our *representation* of our world in natural science, such is essential to the success of our technological activities, and that which we *will* of our world so that it may come to satisfy our aims and desires as viewed from the highest possible level. From this point of view, it is that which is of the essence of social religion that tells us *where* we are going — it is this which tells us of our *destination*, in time — while it is of the nature of science that it can help us on our way to this destination. And then, of course, to the extent that one of these α or ß poles of technology is obscured, so this technology fails. In this scheme of things, the myths of modern science are closer to the ß pole, while, by comparison, the number myth of this present-day technology is closer to the α pole.

In my *magnum opus* I have described this place that is occupied by the technology of the informational revolution, in particular, as a 'middle ground', situated between the grounds of science and religious consciousness. This is however rather precisely the ground that was occupied in earlier times by astrology and alchemy. Today, for by far the greater part, the matters that once concerned these earlier occupiers of the middle ground have passed into darkness, in that they have become occulted, so that they have become the matters of 'the occult'. As such, although they have by no means disappeared from our world — a visit to any popular bookshop should disabuse us of that notion, they have almost entirely lost their earlier significance within society. The task of astrology, of predicting the courses of our outer world and so of guiding these courses to suit our perceived interests, is today the *established* task of technology, while the task of alchemy, of guiding our inner world in the direction of wholeness or 'self realisation', is today increasingly seen as being essential to the *future right application* of technology. Then, very much as in earlier times the planetary system provided the ultimate allegorical medium of the astrologer, and their progression of mineral-chemical reactions provided the ultimate allegorical medium of the alchemists, so today the totality of that which is essential, behind their creations, provides the ultimate allegorical medium of the technologists.[23] Correspondingly, just as the earlier allegorical structures could only attain to their full consistency and perfection of order through number symbolism, so present-day technology can only come fully to presence to the extent that it becomes fully structured and ordered through the use of numerical models — where 'numerical' is to be understood in the general sense introduced earlier in this book. Since the earlier number symbolism served the same *purpose* as is now served by numerical modelling, which is that of bearing that which it serves across the middle ground, so it can be subsumed under the one allegorical type, which is that of *the number myth*.

Having outlined this *proper position in thought* of technology, however, it becomes already clear that just as this technology must make use of the language and thought world of modern science even though it is by no means itself coincident with modern science, so it must also make use of the language and thought world of dogmatic science even though it is in no way coincident with what is central and essential to dogmatic science. It is in fact here, more than in any other place, that it is essential to distinguish between what is at the centre and what is at the periphery. The number myth, even as a current technological myth, is clearly connected to the central myths and thus the central, essential truths of our existence that find their primary representations in the social religions, and which are treated in Christendom in its dogmatic science. Correspondingly, the number myth is obliged to continue to offer its support to the central truths, but it must always know its own place, which is at the periphery. The number myth might, so to say, be *invited* to the centre,

---

[23]Yet another footnote may be allowed in order to exemplify this otherwise rather abstract idea. Two examples may suffice for this purpose. The first of these is that of the *modelling movement*, whereby small-scale physical representations of masterpieces of technology (motor cars, ships, railway locomotives, etc.) are used to recreate something of the aesthetic impact of the original. The second example is that of the *restoration movement*, in which technological masterpieces (pumping engines, aircraft, ships, surveying instruments, complete railway systems, etc.) are lovingly restored, often at very great effort.

as Jung has described in relation to the introduction of triadic constructions in Christian Gospel, but it enters into that place only 'by invitation'. Any confusion about these relative places — not even to consider a juxtaposition of these places — can only lead to further 'esoteric' or whatever other fantasies, of which this world already has quite enough.

It then follows further from this position in thought of technology that it is absolutely essential to the right direction of technology that it holds on most firmly to the religious centre. Otherwise it will become, to revert to Barth's vocabulary, the instrument of nothingness rather than the instrument for overcoming nothingness. The reason for this necessity has been expounded convincingly and in its most general context within Christian theology by Barth (1960, pp. 350, 354, 355) as follows:

> Nothingness does not possess a nature which can be assessed nor an existence which can be discovered by the creature. There is no possible relation between the creature and nothingness. Hence, nothingness cannot be an object of the creature's natural knowledge. It is certainly an objective reality for the creature. The latter exists objectively in encounter with it. But it is disclosed to the creature only as God is revealed to the latter in His critical relationship. The creature knows it only as it knows God in His being and attitude against it. It is an element in the history of the relationship between God and the creature in which God precedes the creature in His acts, thus revealing His will to the creature and informing it about Himself. As this occurs and the creature attains to the truth — the truth about God's purpose and attitude and therefore about itself — through the word of God, the encounter of the creature with true nothingness is also realised and recognised.

> ... Nothingness is absolutely without norm or standard. The explicable conforms to a law, nothingness to none. It is simply aberration, transgression, evil. For this reason it is inexplicable, and can be affirmed only as that which is inherently inimical. For this reason it can be apprehended in its aspect of sin only as guilt, and in its aspect of evil and death only as retribution and misery, but never as a natural process or condition, never as a subject of systematic formulation, even though the system be dialectical. Being hostile before and against God, and also before and against His creature, it is outside the sphere of systematisation. Its defeat can be envisaged only as the purpose

and the end of the history of God's dealings with His creature,
and in no other way.

... God alone can summon, empower and arm the creature to
resist and even to conquer this adversary.

... The creature as such would be no match for nothingness and
certainly unable to overcome it.[24]

Very much again follows from this position in thought of technology when it is
transposed to our outer world of name and form. In the first instance,
technology can only properly *serve* this world through a spirit of dedication; but
dedication is itself provided only by a focusing of devotion. Where devotion
fails it fails everywhere, so that dedication fails too, with all manner of
calamitous consequences. The current environmental disasters of the Soviet
Union and of Central and Eastern Europe are poignant reminders of this, but
any society that places a myth in the wrong place, as 'red mythology' was
placed in the wrong place within these societies, must come to experience
technological disaster. And in this respect science, as and by itself, can provide
no remedy. Thus, just to take one particularly topical example, a nuclear power
plant can only be constructed rapidly and at reasonable cost and can
subsequently only be operated safely and efficiently through dedication, and all
the safety systems, operating manuals, training courses, quality assurance audits,
drills, exercises, simulations and everything else of this kind will not
compensate for a lack of dedication. It follows further, just to keep to this
example, that a society that, in its heart, rejects dedication, and even comes to
despise dedication, so that, deep-down, it despises technology, must experience
great difficulties in building and operating nuclear plant safely and
economically.

Technology has the purpose of creation, and if it is to serve a right
purpose every individual creation must be, to use the singular form of the title
of Kierkegaard's most central work, a *Deed of Love*. Otherwise nothing really
worth while is produced. Although such a direct manner of expression as this

---

[24]Such is the status of these matters today that I cannot help but envisage the prospective
reader of this book flicking through its pages (just as likely backwards as forwards!), coming
across all this talk about God, and literally throwing the thing back from where it came. But I
have to repeat that even those who consider themselves 'in no way religious' have no other
choice than to learn to understand the language and thought world of theistic allegory if they are
to understand the essential problems of our time. And then, just to the extent that their dedication
leads them to make this effort, so they will learn that they are in fact 'religious' after all — and
necessarily so — even though they belong to no visible communion, to no visible church. For
another Christian view of this development, reference should be made to the works of Teilhard
de Chardin. For an Islamic view, reference might be made to Waqar Ahmed Husaini (1980).

is quite out of fashion in our modern societies, so that to speak in this way is almost certain to embarrass the contemporary technologist, yet this person, he or she individually, will none the less recognise that it is basically correct. In any 'sciencewhite' technology, or 'technology of the right hand', or even just a 'post-modern' technology, dedication is essential and can only arise through a focusing of devotion, as a 'deed of love'. Then, on the one hand, it is this same dedication that is at the heart of everything which perseveres and so survives to assert its truth, even if this is only recognised in the mundane and average world in the form of 'business competitiveness', and once again all the science in the world cannot compensate for the absence of dedication: 'Unless the Lord builds the house, its builders work in vain'. On the other hand, but quite consentaneously, it is 'the love of the product' that inspires the contemporary technologist and which focuses such devotion as is present within this individual to dedication, whereby 'the product' itself becomes the allegorical medium for the inner development of the individual technologist, thereby taking on much the same role as did the planetary motions to the astrologer and the mineral-chemical processes to the alchemist. As I have shown in detail in my main work, there is no other way of accounting for the observed realities of technological progress. And indeed, it was shown long ago by writers on political economy as diverse as Rosa Luxemburg and Maynard Keynes, and already for the case of the industrial revolution, that such 'technology-led' revolutions can never be explained by such gross agents as 'the profit motive', but that they possess, as it were, 'a life of their own', driven by acts of apparently quite irrational dedication. Such agents as the profit motive certainly act as means, but they cannot in practice serve as ends.

It remains further to add that although technology is properly strung between science and the religious consciousness, it can in no way be associated with any particular social religion. In fact, concerning this often vexing problem of the relation of technology to a specific social-religious environment or culture (as was introduced by Max Weber at the beginning of this century in his doctrine of the 'Protestant work ethic'), it should be observed that, corresponding to the middle-ground position of technology, situated below that of the 'high ground' of that which is essential to social religion, the predominant relation between technology and social religion can only be one of a *falling away* from social religion. For example, our present era of informational revolution was prepared to a numerically disproportionate extent by Jews or persons of predominantly Jewish parentage (*e.g.* Cantor, Einstein, Freud, Husserl, Kafka, Mahler and Wittgenstein), but these were nearly all Jews who had fallen away, to a greater or lesser degree, from the 'high ground' of Judaism, even if (as in the cases of Husserl and Mahler) they adopted other faiths. (This same point was also made by Britain's chief rabbi-elect, Dr. Jonathan Sachs, in his 1990 BBC Reith lecture: in fact the only exception of which I can think, out of hand, is that of Martin Bubner.) Thus although Judaism as such had nothing whatsoever to say that was specific to the first stages of the informational revolution, any more than Protestantism as such had anything specific to do with the industrial revolution, yet the falling away from the respective specific cultural environments clearly did provide social-economic advantages. We observe similarly how, as the informational revolution has continued, its progress has been most strongly promoted in the East by individuals falling away from Buddhism, and indeed we are today witnesses of some individuals in the West who appear to approach Buddhism precisely for the social advantages that its practices provide, even as they have, in their hearts, no concern for Buddhist enlightenment, for Buddhism as such, so that they approach Buddhism specifically in order to fall away from it. But then, from the point of view of our current informational societies, it is only in this sense of a falling-away that there can be any immediate and explicit relation between social-economic advantage and social religion.

Once again, in concluding, I have to apologise for such a superficial, fragmentary and unsupported exposition of these matters, which I have treated in detail and at depth in my main work, but I hope that this may suffice for the attentive reader to get a first notion of the proper position of technology within society, and thus of the proper place of hydroinformatics also. I hope that I may at least have suggested to the reader that, without its connection to the religious, which is something quite other than the natural scientific, hydroinformatics can easily come to offer extraordinary opportunities only for a further and even more ruthless misuse of the resources of our planet, and thus can have only further 'world-catastrophic consequences'. On the other hand, if hydroinformatics is aware of its proper place and so allows itself to be guided by what humanity has come to learn about the religious, it can come to contribute mightily to saving our world from environmental cataclysm. For the purposes of guiding it in this way, attention must be given to the less conventional matters introduced in this book and emphasised in this concluding unscientific postscript.

References

Abbott M.B., 1959, Structural analysis by digital computer, *Engineering 187,* No. 4863, pp. 666-667.

Abbott M.B., 1979, *Computational hydraulics, elements of the theory of free-surface flows*, Pitman/Longman, London.

Abbott M.B., 1986, Computer modelling: a warning notice. *The Dock and Harbour Authority*, March, pp. 251-256.

Abbott M.B., 1989, Modelling of the coastal environment, in Falconer R.A., Goodwin P., Matthew R.G.S. (editors), *Proceedings of the Int. Conf. on Hydraulic and Environmental Modelling of Coastal, Estuarine and River Waters,* Gower, Aldershot.

Abbott M.B., 1990(a), Contributions of computational hydraulics to the foundation of a computational hydrology, in Bowles D.S. and O'Connell P.E. (editors), *Recent advances in the modelling of hydrologic systems*, Kluwer, Amsterdam.

Abbott M.B., 1990(b), The impact of ESPRIT projects upon the modelling of hydrologic systems, in Bowles D.S. and O'Connell P.E. (editors), *Recent advances in the modelling of hydrologic systems*, Kluwer, Amsterdam.

Abbott M.B., 1991(a), Numerical modelling for coastal and ocean engineering, in Herbich J.B. (editor), *The handbook of ocean and coastal engineering,* Gulf, Houston.

Abbott M.B., 1991(b), A central issue of teaching within the hydroinformatics paradigm, Proc. Colloquium Société Hydrotechnique de France: la Formation de l'Ingénieur Hydraulicien Européen de l'an 2000, to appear in *La Houille Blanche*, 1991.

Abbott M.B. and Basco D.R., 1989, *Computational fluid dynamics, an introduction for engineers*, Longman, London, and Wiley, New York.

Abbott M.B. and Cunge J.A. (editors), 1981, *Engineering applications of computational hydraulics: homage to Alexandre Preissmann*, Pitman, London.

Abbott M.B. and Madsen P.E., 1990, Modelling of wave agitation in harbours, in Hanes D.M. and Le Méhauté, B. (editors), *The Sea, Vol. 9*, Wiley, New York.

Abbott M.B. and Price W.A. (editors), 1992, *The coastal, estuarial and harbour engineer's reference book*, Chapman and Hall, London.

Abbott M.B. and Warren I.R., 1974, A dynamic population model, *J. Env. Management, 2*, pp. 284-297.

Abbott M.B., Damsgaard Aa. and Rodenhuis G.S., 1973, System 21, Jupiter, a design system for two-dimensional nearly-horizontal flows, *J. Hydraulic Res. 11*, pp. 1-28.

Abbott M.B., Hodgins D.O., Dinsmore A.F. and Donovan M., 1977, A transit model for the city of Miami, *J. Env. Management, 5*, pp. 229-242.

Abbott M.B., de Nordwall J. and Swets B., 1983, On applications of artificial intelligence to control and safety problems of nuclear power plants, *Civ. Eng. Syst. 1*, pp. 69-82.

Abbott M.B., Bathhurst J.C., Cunge J.A., O'Connell P.E. and Rasmussen J., 1985, An introduction to the European Hydrologic System (SHE), *J. Hydrology 87*, pp. 45-59 and 66-77.

Abbott M.B., Bardis L., Hornsby C.P.W., Katsoulakos P.S., Lind M. and Wittig T., 1988, An architecture for a shipboard knowledge-based system, *ESPRIT 88, Putting the technology to use, Part 1*, North Holland, Amsterdam, pp. 780-795.

Abbott M.B., Havnø K. and Lindberg S., 1991, The fourth generation of numerical modelling in hydraulics, to appear in *J. Hydraulic Res.*

ACM, 1977, ACM-SIGPLAN/SIGART Proc. ACM/Symposium on artificial intelligence and programming languages, *SIGPLAN Notices, 12-8 Symp.*, and *SIGART Newsletter 64*.

ACM, 1981, ACM-SIGART, SIGMOD, SIGPLAN Proc. Workshop on data abstraction, databases and conceptual modelling, *SIGART Newsletter 74*.

Azam F., Fenchal T., Field J.G., Grey I.A., Meyer-Reil I.A. and Thingstad F., 1983, The ecological role of water column microbes in the sea, *Mar. ecol. new series 10*, pp. 257-263.

Baretta J.W. and Ruardij P. (editors), 1988, *Tidal flat estuaries, simulation and analysis of the Ems estuary*, Springer, Heidelberg.

Barnes R.S.K. and Hughes R.N., 1982, *An introduction to marine ecology*, Blackwell, Oxford.

Baroody A.J. and de Witt D.J., 1981, An object-oriented approach to data base system implementation, *ACM translation on data base systems 6-4*, pp. 576-601.

Barr A. and Feigenbaum E.A., 1981, *The handbook of artificial intelligence, Vol. 1*, Pitman, London.

Barth K., 1922/1931, *The epistle to the Romans*, translation Hoskyns E.C., Oxford Univ. Press, Oxford.

Barth K., 1960, *Church dogmatics, Vol. III: the doctrine of creation, Part 3*, translation Bromiley G.W. and Ehrlich R.J., Clark, Edinburgh.

Biemel W., 1973/1977, *Martin Heidegger, An illustrated study*, translation Mehta J.L., Routledge and Kegan Paul, London.

Bird R. and Wadler P., 1988, *Introduction to functional programming*, Prentice Hall, New York.

Bohm D., 1980, *Wholeness and the implicate order*, Routledge and Kegan Paul, London.

Browder F.E. (editor), 1976, *Proc. symp. pure math., XXVIII*, American Mathematical Society, Providence, Rhode Island.

Capra F., 1978, *Tao of physics*, Fontana, New York.

Carnap R., 1937, *Foundations. of logic and mathematics*, Univ. Chicago Press, Chicago.

Carnap R., 1950, *Logical foundations of probability*, Univ. Chicago Press, Chicago.

Carnap R., 1956, *Meaning and necessity: a study in semantics and modal logic* (fourth edition), Univ. Chicago Press, Chicago.

CEC, 1988, *Esprit 88, putting the technology to use*, North Holland, Amsterdam.

Chaitin C.J., 1982, Gödel's theorem and information, *Int. J. Theoretical Phys.* *22*, p. 941-954.

Chaitin C.J., 1987, Incompleteness theorems for random reals, *Advances in applied math. 8*, pp. 119-146.

Chorin A.J., 1983, Review of Peyret R. and Taylor T.D., Computational methods for fluid flow, *Bull. Am. Math. Soc. 9* (3), pp. 368-372.

Cooke D.J. and Bez H.E., 1984, *Computer Mathematics*, Cambridge Univ. Press, Cambridge.

Cunge J.A., 1989, Review of recent developments in river modelling, in Falconer R.A., Goodwin P. and Matthew R.G.S., *Proc. int. conf. hydraulic and environmental modelling of coastal, estuarial and river waters*, Gower, Aldershot, pp. 393-410.

Cunge J.A., 1991, Informatique, hydraulique numérique, hydroinformatique, Proc. Colloquium Société Hydrotechnique de France: La Formation de l'Ingénieur Hydraulicien Européen de l'an 2000, to appear in *La Houille Blanche, 1991*.

Cunge J.A., Holly F.M. and Verwey A., 1980, *Practical aspects of computational river hydraulics*, Pitman, London and Iowa Univ. Press, Iowa.

Dahl O.J., Dijkstra E.W. and Hoare C.A.R., 1972, *Structural programming*, Academic, New York.

Date C.J., 1989, *A guide to the SQL standard: a user's guide to the standard relational language SQL* (second edition), Addison-Wesley, Reading M.A.

Davis R., 1980, Meta-rules: reasoning about control, *Artificial Intelligence*, 15.

De Coursey D.G. (editor), 1990, *Proceedings of the international symposium on water-quality modelling of agricultural non-point sources*, 2 volumes, U.S. Department of Agriculture, Washington D.C.

Derrida R., 1972/1982, *Margins of philosophy*, translation Bass A., Harvester, London.

DHI, 1990, *MIKE 21 short description*, Danish Hydraulics Institute, Hørsholm, Denmark.

DHI, 1990, *Environmental Monitoring Systems - The Great Belt - A Case Story*. Danish Hydraulic Institute, Hørsholm.

Diamond C. (editor), 1976, *Wittgenstein's lectures on the foundations of mathematics, Cambridge, 1939*, Harvester, Hassocks (Sussex, UK).

Dilthey W., 1976, *Selected writings*, translation Rickman H.P., Blackwell, London.

Dijkstra E.W., 1972, *The humble programmer*, CACM 13-10, pp. 839-866.

Eckmiller R. (editor), 1990, *Advanced neural computers*, North Holland, Amsterdam.

Engelmore R. and Morgan J., 1988, *Blackboard Systems*, Addison-Wesley, Wokingham.

Fenchal T., 1988, Marine plankton food chains, *Ann. rev. ecol. 19,* pp. 19-38.

de Finetti B., 1974, *Theory of probability*, translation Machi A. and Smith A., Wiley, New York.

Finkelstein D., 1977, Beneath time: explanation in quantum topology, in Fraser J.T., *et al* (editors), *The study of time III*, pp. 94-114.

Fishman K.D., 1981, *The computer establishment*, Harper and Row, New York.

France A., 1923, *The garden of Epicurus*, translation Allison A., Dodd-Mead, New York, pp. 213-214.

Genesereth M.R. and Nilsson N.J., 1988, *Logical foundations of artificial intelligence*, Morgan, Kaufmann, Palo Alto.

Gilson E., 1950, *History of Christian philosophy in the middle ages*, Sheed and Ward, London.

Gofuku J., Yoshikawa H., Itoh K., Wakabayashi J., 1986, The study of a time critical diagnostic method for emerging operation of nuclear power plant, *Proc. 6th power plant dynamics, control and testing symposium,* Univ. Tennessee, pp. 63-87.

Habermas J., 1968/1972, *Knowledge and human interests*, Heinemann, London.

Halfon E., 1983, Is there a best model structure? II Comparing the model structures of different fate models, *Ecological Modelling 20*, pp. 155-163.

Harmon P. and King D., 1985, *Expert systems: artificial intelligence in business*, Wiley, New York.

Harmon P. and Mans R., 1988, *Expert systems: tools and applications,* Wiley, New York.

Harrison, W. and Stevens, C.F., 1976, Bayesian Forecasting. Lecture read before the Royal Statistical Society.

Hartree D.R., 1952, *Numerical analysis,* Oxford Univ. Press, Oxford.

Hayes-Roth B.I., 1985, A blackboard architecture for control, *Artificial Intelligence 26,* pp. 251-321.

Hayes-Roth F., Waterman D.A. and Lenat D.B. (editors), 1985, *Building expert systems,* Addison-Wesley, Reading, Mass.

Heemink A.W., 1986, Storm surge prediction using Kalman filtering, *Rijkswaterstaat Commun. 46,* The Hague.

Heemink A.W. and Kloosterhuis H., 1990, Data assimilation for non-linear tidal models, *Int. J. Non-linear Methods in Fluids 11,* pp. 1097-1112.

Heidegger M., 1961, *Der europäische Nihilismus,* Neske, Pfullinger.

Heidegger M., 1927/1962, *Being and time,* translation Macquarrie J. and Robinson E., Blackwell, London.

Heidegger M., 1969, *Identity and difference,* translation Stambaugh J., Harper and Row, New York.

Heidegger M., 1977, *The question concerning technology, and other essays,* translation Lovitt W., Harper and Row, New York.

van Heijenoort J., 1967, editor, *From Frege to Gödel: a source book in mathematical logic,* Harvard Univ. Press, Cambridge, Mass.

Heppenheimer T.A., 1990, 'The long shadow of Kurt Gödel, *MOSAIC 21, 1,* National Science Foundation, Washington D.C.

Hitchcock E.A., 1857/1977, *Alchemy and the alchemists,* Phil. Res. Soc., Los Angeles.

Hoogstraten H.J., 1985, *Irrigation and social organisation in West Malaysia,* Dept. of Sociology, Agr. Univ., Washington.

Hoyle F., 1980, *The relation of biology to astronomy,* U.C. Cardiff, Cardiff, UK.

Huntley H.E., 1970, *The divine proportion,* Dover, New York.

Husserl E., 1900, 1913/1970, *Logical investigations,* translation Findley J.N., Routledge and Kegan Paul, London.

Husserl E., 1938/1973, *Experience and judgement, investigations in a genealogy of logic,* translation Churchill J.S. and Ameriks K., Routledge and Kegan Paul, London.

IRC, 1988, *Community participation and women's involvement in water supply and sanitation projects,* Int. Ref. Centre for Community Water Supply and Sanitation, The Hague.

Itai A. and Roden M., 1984, The multi-tree approach to reliability in distributed networks, *Proc. 25th Symp. on Foundations of Computer Science,* pp. 137-147.

Jaspers K. and Bultmann R., 1954, *Die Frage der Entmythologisierung,* Piper, Munich.

Jeffrey R., 1984, de Finetti's probabilism, *Synthese, 60,* pp. 70-90.

Joint I.R. and Morris R.J., 1982, The role of bacteria in the turnover of organic matter in the sea, *Oceanogr. Mar. biol. A Rev. 20,* pp. 65-118.

Jørgensen S.E., 1986, *Fundamentals of ecological modelling,* Elsevier, Amsterdam.

Jung C.G., 1944/1953, *Psychology and alchemy,* translation Hall R.F.C., Routledge and Kegan Paul, London.

Jung C.G., 1943/1953, *Collected works* (edited by Read H., Fordam M. and Adler G.; translation Hall R.F.C.), Routledge and Kegan Paul, London.

Jung C.G., 1952/1955, *Synchronicity, an acausal connecting principle,* translation Hall R.F.C., Routledge and Kegan Paul, London.

Kaviratna H., 1980, *Dhammapāda* (translation and annotation), Theos. Univ. Press, Pasadena.

Kay A., 1984, Inventing the future, in Winston H. and Prendergast K.A. (editors), *The AI business: the commercial uses of artificial intelligence.* MIT, Cambridge.

Kiefer D.A. and Kiefer J.N., 1987, Origins of vertical patterns of phytoplankton and nutrients in the temperate open ocean: a stratigraphic hypothesis *DSR 28A,* pp. 1087-1105.

Kierkegaard S.Aa., 1847, *Kjærlighedens Gjerninger (Deeds of love).*

Kierkegaard S.Aa., 1849, *Sygdommen til Døden (The sickness unto death).*[25]

Kierkegaard S.Aa., 1859, *Synspunktet for min Forfatter-Virksomhed, En ligefrem Meddelelse, Rapport til Historien (Aspects of my activity as an author: a straightforward announcement; a report to history).*

Kirkegaard J., Mogensen B. and Vested H.J., 1991, A coupled hydrographic monitoring and forecasting system for the Great Belt, *Am. Met. Soc., 71st Annual Meeting,* New Orleans.

Kleene S.C., 1977, Foundations of mathematics, *Encyclopædia Britannica, 15th edition,* 11.

Kolmogorov A.N. and Fomin S.V., 1957, *Elements of the theory of functions and functional analysis,* translation Boron L.F., 2 Vols., Greylock, New York.

Kowalik J.S. (editor), 1986, *Coupling symbolic and numerical computing in expert systems,* North Holland, Amsterdam.

Kreisel C., 1976, What have we learnt from Hilbert's second problem, in Browder F.E. (above), pp. 93-130.

---

[25]English translations in this case include editions from Penguin and Princeton University Press. There are, of course, translations of almost all Kierkegaard's works into almost all languages.

Kremer J.N. and Nixon S.W., 1978, A coastal marine ecosystem simulation and analysis, *Ecological studies 24*, Springer, Heidelberg.

Lakatos I., 1976, *Proofs and refutations: the logic of mathematical discovery* (edited by Worrall J. and Zahar G.), Cambridge Univ. Press, Cambridge.

Lamb H., 1931, *Hydrodynamics*, Cambridge Univ. Press, Cambridge.

Leendertse J.J., 1981, Discussion, in Fischer H.B. (editor), *Transient models for inland and coastal waters*, Academic Press, New York, pp. 31-35.

Leslie D.C. and Quarini G.L., 1979, The application of turbulence theory to the formulation of subgrid modelling procedures, *J. Fluid Mech., 97*, pp. 65-91.

Lévi-Strauss C., 1958, *Anthropologie structurale*, Plon., Paris.

Lévi-Strauss C., 1962, *La pensée sauvage*, Plon., Paris.

Lightstone A.G., 1978, *Mathematical logic, an introduction to model theory*, Plenum, New York.

Liskov B.H., 1974, A note on CLU, *Project MAC, Memo 112*, MIT, Cambridge, MA.

Liusternik L. and Sobolev V., 1961, *Elements of functional analysis*, translation Labarre A.E., Izbicki H. and Crowley H.W., Ungar, 1979, New York.

Lockman P.C., Shaw M. and Wulf W.A., 1979, Data abstraction for database systems, *ACM translation on data base systems 4-1* pp. 60-75.

Lorentz C., 1989, Quality as a competitive commodity, *Financial Times*, 13 Nov.

Lowenthal E.I., 1971, Structures for generalised data management systems, Ph.D. dissertation, Dept. Comp. Sc. Univ. Texas, Austin.

Lyndon H.C., 1964, *Notes on logic*, van Nostrand, New York.

McCarthy J., 1979, Ascribing mental qualities to machines, *Tech. report STAN-CS-79-725*, AIM-326, Stanford Univ. Press.

Manin Yu I., 1977, *A course in mathematical logic*, Springer, New York.

Mann K.H., 1982, *Ecology of coastal waters, a systems approach*, Blackwell, Oxford.

Margalef R., 1968, *Perspectives in ecological theory*, Univ. of Chicago Press, Chicago.

Melkanoff M.A. and Zamfir M., 1978, The automatization of data base conceptual models by abstract data types, Comp. Language Group, *Comp. Sc. Report UCLA-ENG-7785*, UCLA, Los Angeles.

Mendelson E., 1964, *Introduction to mathematical logic*, van Nostrand, New York.

Meyer B., 1988, *Object-oriented software construction*, Prentice Hall, New York.

Milne Thomsen L.M., 1955, *Theoretical hydrodynamics*, Macmillan, London.

Møller, J.S., 1989, Denmark's Great Belt Link, Environmental and Hydrodynamic Issues. Paper presented at ASCE Annual Convention, New Orleans, October.

Morgado E.J.M., 1986, *Semantic networks as abstract data types*, Dept. Comp. Sc., Univ. New York, Buffalo.

Moss B., 1988, *Ecology of fresh waters, man and medium* (2nd edition) Blackwell, London.

Nietzsche F., 1887/1969, *Die fröhliche Wissenschaft* (*la Gaya Scienza*) Ullstein, Frankfurt/M.

Nilsson N.J., 1980, *Principles of artificial intelligence*, Morgan Kaufmann, Palo Alto.

Ormsbee L. and Kessler A., 1990, Optimal upgrading of hydraulic network reliability, *J. Water Resources Planning and Manag.*, ASCE, 116, pp. 784-802.

Penrose R., 1989, *The Emperor's new mind*, Cambridge Univ. Press, Cambridge.

Poincaré H., 1900±, *Science and method*, Dover, New York.

Popper K.R., 1936/1959, *The logic of scientific discovery*, Hutchinson, London.

Popper K.R., 1963, *Conjectures and refutations*, Routledge and Kegan Paul, London.

Quirk R. and Greenbaum S., 1973, *A university grammar of English*, Longman, London.

Radach G. and Moll A., 1989, State of the art in algal bloom modelling, EEC workshop on eutrophication and algal blooms in North Sea coastal zones, the Baltic and adjacent areas, *Water Pollution Research Reports*.

Radford P.J., 1971, The simulation language as an aid to ecological modelling, in Jeffers J.N.R., *Math. models in ecology*, Proc. Bristol Ecological Soc., pp. 277-296.

Radford P.J., 1979, Some aspects of estuarine ecosystem model GEMBASE-state of the art in ecological modelling, in Jørgensen S.E., *Proceedings of the ISEM conference on ecological modelling, Copenhagen, Int. Soc. for ecological modelling*, pp. 301-322.

Rajagopalachari C., 1952, *Rāmāyaṇa* (translation and annotation), Bhavan, Bombay.

Richtmyer R.D., 1957, *Difference methods for initial value problems*, Interscience, New York.

Riesz F. and Sz Nagy B., 1955, *Functional analysis*, translation Boron L.F., Ungar, New York.

Roman P., 1975, *Some modern mathematics for physicists and other outsiders*, Pergamon, New York.

Rumelhart D., McClelland J. and the PDP research group, 1987/1988, *Parallel distributed processing*, MIT Press, Boston.

Schleiermacher F., 1831/1976, *The Christian faith*, translation Mackintosh H.R. and Stewart J.S., Clark, Edinburgh.

Schramm G. and Warford J.J. (editors), 1989, *Environmental management and economic developments*, World Bank, Washington D.C.

Shackle G.L.S., 1967, *The years of high theory: invention and tradition in economic thought, 1926-1939*, Cambridge Univ. Press, Cambridge.

Shaw M., 1980, Abstraction, data types and models for software *ACM 81*, pp. 189-191.

Shaw R.R. and Falco J.W., 1990, A management perspective on water-quality models for agricultural non-point sources of pollution, in De Coursey D.G. (editor).

Smagorinski J., 1963, General circulation experiments with the primitive equations, *Monthly Weather Rev.*, NWB, *93*:99.

Smetacek V. and Pollehne F., 1986, Nutrient cycling in pelagic systems: a reappraisal of the conceptual framework, *Ophelia 26*, pp. 401-428.

Smith J.M. and Smith D.C.P., 1977, Data base abstractions: aggregation, *CACM 20*, pp. 405-413.

Steele L., 1991, *Lectures on knowledge systems*, Addison Wesley, New York.

Stefik M. and Bobrow D., 1986, Object-oriented programming: themes and variations, *AI Magazine 6* (4), pp. 40-64.

Steiner G., 1978, *Heidegger,* Fontana, London.

Stigebrand A. and Wulff F., 1987, A model for dynamics of nutrients and oxygen in the Baltic proper, *J. Mar. Res. 4S*, pp. 729-757.

Sutram A., 1972, *La cabale*, Puyot, Paris.

Thayse A. (editor), 1988, *From standard logic to logic programming*, Wiley, New York.

Valéry P., 1957-1960, *Œuvres*, Gallimard, Paris, Vol. I, p. 202.

Valéry P., 1960-1973, *Collected Works*, translation Mathews J., Bollinger Foundation, New York, Vol. 1, p. 251.

Vested, H.J., Jensen, H.R., Petersen, H.M., Jørgensen, A.M. and Machenhauer, B., 1990, *An Operational Hydrographic Warning System for the North Sea and Danish Belts*. Paper presented at the 1990 JONSMOD Conference, Bidston, U.K.

de Vries I., Hopstraten H., Goosens M., de Vries M., de Vries H., Heringa J., 1988, GREWAQ: an ecological model for lake Grevelingen, *Documentation report T 0215-03*, Delft Hydraulics, Delft.

Wahl J., 1959/1969, Philosophies of existence, translation Lory F.M., Routledge and Kegan Paul, London.

Wakabayashi J., Gofuku A., Okazaki F., Tashima S.I., 1985, Application of projective operator technique for the disturbance identification of nuclear power plants, *Proc. ANS topical Meeting on computer application for nuclear power plant operation and control*, Tri Cities, Washington, pp. 419-426.

Waqar Ahmed Husaini S., 1980, *Islamic environmental systems engineering*, Macmillan, London.

Wetzel R.G., 1983, *Limnology* (2nd edition), Saunders, Philadelphia.

White A. and Gordon G., 1987, *Training community motivators in water supply and sanitation*, Int. Ref. Centre for Community Water Supply and Sanitation, The Hague.

Whitham G.B., 1974, *Linear and non-linear waves*, Wiley, New York.

Winston P.H., 1977, *Artificial intelligence*, Addison-Wesley, Reading, MA.

Wittgenstein L., 1969, *Schriften IV, Philosophische Grammatik, Teil II, Über Logik und Mathematik*, Suhr, Frankfurt.

Wittgenstein L., 1975, *Philosophical Remarks*, translation Hargreaves R. and White R., Blackwell, Oxford.